A Research Agenda for Climate Justice

Elgar Research Agendas outline the future of research in a given area. Leading scholars are given the space to explore their subject in provocative ways, and map out the potential directions of travel. They are relevant but also visionary.

Forward-looking and innovative, Elgar Research Agendas are an essential resource for PhD students, scholars and anybody who wants to be at the forefront of research.

Titles in the series include:

A Research Agenda for Entrepreneurial Cognition and Intention
Edited by Malin Brännback and Alan L. Carsrud

A Research Agenda for Entrepreneurship Education
Edited by Alain Fayolle

A Research Agenda for Service Innovation
Edited by Faïz Gallouj and Faridah Djellal

A Research Agenda for Global Environmental Politics
Edited by Peter Dauvergne and Justin Alger

A Research Agenda for New Institutional Economics
Edited by Claude Ménard and Mary M. Shirley

A Research Agenda for Regeneration Economies
Reading City-Regions
Edited by John R. Bryson, Lauren Andres and Rachel Mulhall

A Research Agenda for Cultural Economics
Edited by Samuel Cameron

A Research Agenda for Environmental Management
Edited by Kathleen E. Halvorsen, Chelsea Schelly, Robert M. Handler, Erin C. Pischke and Jessie L. Knowlton

A Research Agenda for Creative Tourism
Edited by Nancy Duxbury and Greg Richards

A Research Agenda for Public Administration
Edited by Andrew Massey

A Research Agenda for Tourism Geographies
Edited by Dieter K. Müller

A Research Agenda for Economic Psychology
Edited by Katharina Gangl and Erich Kirchler

A Research Agenda for Entrepreneurship and Innovation
Edited by David B. Audretsch, Erik E. Lehmann and Albert N. Link

A Research Agenda for Global Crime
Edited by Tim Hall and Vincenzo Scalia

Research Agenda for Transport Policy
Edited by John Stanley and David A. Hensher

A Research Agenda for Tourism and Development
Edited by Richard Sharpley and David Harrison

A Research Agenda for Financial Inclusion and Microfinance
Edited by Marek Hudon, Marc Labie and Ariane Szafarz

A Research Agenda for Climate Justice
Edited by Paul G. Harris

A Research Agenda for Climate Justice

Edited by

PAUL G. HARRIS

Chair Professor, Global and Environmental Studies, The Education University of Hong Kong and Senior Research Fellow, Earth System Governance global research alliance

Elgar Research Agendas

Edward Elgar
PUBLISHING

Cheltenham, UK • Northampton, MA, USA

Published by
Edward Elgar Publishing Limited
The Lypiatts
15 Lansdown Road
Cheltenham
Glos GL50 2JA
UK

Edward Elgar Publishing, Inc.
William Pratt House
9 Dewey Court
Northampton
Massachusetts 01060
USA

Paperback edition 2021

A catalogue record for this book
is available from the British Library

Library of Congress Control Number: 2019951008

This book is available electronically in the **Elgar**online
Social and Political Science subject collection
DOI 10.4337/9781788118170

ISBN 978 1 78811 816 3 (cased)
ISBN 978 1 78811 817 0 (eBook)
ISBN 978 1 80088 839 5 (paperback)

Typeset by Servis Filmsetting Ltd, Stockport, Cheshire
Printed and bound by CPI Group (UK) Ltd, Croydon, CR0 4YY

Contents

Contributors

Robin Attfield is Professor Emeritus in the School of English, Communication and Philosophy, and in the Institute for Sustainable Places, at Cardiff University, UK.

Ian Bailey is Professor of Environmental Politics in the School of Geography, Earth and Environmental Sciences at the University of Plymouth, UK.

Fausto Corvino is a postdoctoral fellow in Moral Philosophy at the Sant'Anna School of Advanced Studies, Pisa, Italy.

Alix Dietzel is Lecturer in Global Ethics in the School of Sociology, Politics and International Studies at the University of Bristol, UK.

Justin Donhauser is a Junior Faculty member in the Philosophy Department at Bowling Green State University, USA.

Paul G. Harris is the Chair Professor of Global and Environmental Studies at the Education University of Hong Kong.

Sanna Kopra is a postdoctoral fellow in the Arctic Centre at the University of Lapland, Finland.

James S. Mastaler is Lecturer in Theology at Loyola University Chicago, USA.

Selina Rose O'Doherty is Lecturer in International Relations at the School of History, Anthropology, Philosophy and Politics at Queen's University, Belfast, UK.

Giuseppe Pellegrini-Masini is a postdoctoral research fellow in the Department of Psychology at NTNU Norwegian University of Science and Technology, Trondheim.

Alberto Pirni is Senior Researcher and Lecturer in Moral Philosophy at the Sant'Anna School of Advanced Studies, Pisa, Italy.

David E. Storey is Associate Professor of the Practice in the Philosophy Department at Boston College, USA.

Claire Swingle is a Princeton in Asia Fellow at the International Water Management Institute in Kathmandu, Nepal.

Carlos Tornel is Associate Professor in the Department of International Studies and a member of the Interdisciplinary Center for Sustainability at Universidad Iberoamericana, Mexico City.

Ivo Wallimann-Helmer is Assistant Professor for Environmental Humanities at the University of Fribourg, Switzerland.

Preface

During my three decades of studying the politics, policy and ethics of climate change, relatively little progress has been made in realizing climate justice. Indeed, the problem has only gotten worse. Greenhouse gas pollution continues to increase globally. The impacts of climate change are increasingly being felt around the world. It is all but certain that they will accelerate and widen with time. So, too, will the resulting suffering to individuals, communities and other species. Meanwhile, most of the world's affluent people and communities are living high on the hog, using energy and consuming stuff as if there were no tomorrow – and as if the consequences for the world's poor and weak did not matter. In the context of climate change, injustice far outstrips justice.

What should we do about this? The answer is long and complex, so it may be simpler to put the question in negative terms: what should we *not* do about this? Put this way, the answer becomes simpler: we – every capable person and every other capable actor (governments, businesses and all the rest) – should not continue doing what we are doing now. If we do not change our ways dramatically and quickly, climate change will definitely be the greatest injustice experienced in all of human history. The longer we delay, the greater the injustices to come.

As part of efforts by scholars and activists around the world to mitigate both the causes of climate change and the injustices it portends, this book aims to portray and stimulate alternative, innovative perspectives for understanding climate justice (through theorizing and researching) and implementing it (through individual behaviour changes and collective policy actions). It comprises of a dozen original chapters by scholars who are more than disinterested researchers. By my reckoning, each of them shares my frustration with the lack of robust global action on climate change. Each has a genuine desire to further the realization of climate justice. With luck, their ideas may help to nudge the world toward that objective, or at least move it a bit away from so much climate injustice.

Every book is a collaboration. This one is no exception. My thanks go to Rachel Downie at Edward Elgar Publishing for asking me to put this collection together. I trust that it will make a valuable addition to Elgar's list of books on climate change. I am obliged to the contributors for surrendering to my arm twisting by writing for this book. My appreciation goes to anonymous referees who read and commented on early drafts of the chapters. Finally, for companionship and domestic support during this project and so many others over the years, I am grateful to Keith, Mobie and especially our beloved Susie, who was at my side during all of my work on this book apart from the last few days, when she sadly passed away. Susie's passing was absolutely terrible, but there was nothing unjust about it. That is because it was nobody's fault. We cannot say the same about climate change.

Paul G. Harris
Lantau Island
September 2019
www.paulgharris.net

1   Climate justice: the urgent research agenda(s)

Paul G. Harris

Climate change has moved rapidly from the fringes of scientific research to become a topic that is central to debates among scholars from a wide variety of disciplines. It is now a front-burner issue for governments, businesses, civil-society organizations and publics. As the science of climate change has improved, and as its impacts have been increasingly felt around the world, it has become clearer that this issue is about much more than changes to the natural environment. It is also about politics and the distribution of scarce resources among and within national communities. It is about harsh impacts being felt by some more than others, not least by those individuals and communities least responsible for causing the problem. It is about the disproportionate responsibility of the world's affluent people and nations. In short, climate change is a matter of equity, fairness and social and distributive justice. Addressing climate change squarely requires research that enables a complete comprehension of the importance and implications of justice.

This book aims to advance a research agenda for climate justice that can help to illuminate new pathways forward for scholars (including advanced research students), policymakers, diplomats, practitioners and activists. A key objective of the book is to showcase and stimulate innovative, alternative perspectives on climate justice. Each chapter challenges assumptions and prevailing practices, thereby showcasing potential new directions for research and action on climate justice. In addition to furthering climate justice as a scholarly field, the book aims to have real-world impact by sharing new ideas that will inform and guide those doing the actual work of realizing it. As an introduction to subsequent chapters, in this one I very briefly introduce the problem of climate change and related policies, delineate the concept of climate justice, highlight some other books that have explored this topic recently, and summarize the contributors' individual agendas and arguments.

Climate change and the weakness of climate policy

Scientists first hypothesized in the nineteenth century that warming of the atmosphere would result from so-called greenhouse gases, but it was not until the 1980s that climate change started to receive widespread attention by both scientists and governments (cf. Harris 2019: 4–5). Since then, understanding of the issue has accelerated, with the latest research confirming many people's worst fears: massive anthropogenic emissions of greenhouse gases from human activities, especially global emissions of carbon dioxide from the combustion of fossil fuels, are changing Earth's climate system in profound ways (IPCC 2013). The atmosphere is growing warmer, as are its oceans, leading to myriad negative impacts, including more severe weather events, sea-level rise, spread of disease vectors, losses in agricultural productivity and accelerating extinctions. Without very robust action by governments and other actors to reduce greenhouse gas emissions drastically – more or less bringing net global carbon emissions to a halt within little more than a generation – the impacts will be dangerous and severe. Indeed, even with action it is possible that global warming and other manifestations of climate change will increase rapidly, in part due to 'natural' releases of carbon dioxide and other greenhouse gases from changing ecosystems (e.g. releases from melting permafrost and warming boreal forests).

The real dangers of climate change in the very near future, and the impacts predicted for coming decades and beyond, have started to stimulate action by governments and other actors to mitigate greenhouse gas pollution and grapple with the impacts that are being felt already and will be much more so in the future. At the international level, states (variously referred to in this book as countries, nations and states) came together in the late 1980s and early 1990s to negotiate the United Nations Framework Convention on Climate Change (UNFCCC), an overarching agreement for state-led action that has as its core objective the prevention of 'dangerous anthropogenic interference with the climate system' (UNFCCC 1992: 9). Subsequent agreements, such as the 1997 Kyoto Protocol to the UNFCCC and the 2015 Paris Agreement on climate change, have resulted from almost continuous international negotiations and more or less yearly 'conferences of the parties' to the convention and related agreements (for a summary, see Harris 2018: 129–36). Very early in the climate change negotiations, questions and concerns about equity and justice, particularly international justice among negotiating states, arose and took on prominence (Harris 2001: 44–88). States agreed that their responses to climate change ought to be equitable, fair and just – although agreement on what this means in practice has never been fully determined. The closest to a consensus on the way forward was the principle of 'common but differentiated responsibility', whereby developed states accepted greater responsibility for causing and dealing with climate change. However, while this principle has been an accepted part of international climate negotiations for decades, it is typically practised in the breach. Few disagree that the richest states have greater responsibility – some would say that this is so because they contributed the most to causing climate change, while others would add that they also have special responsibilities due to their affluence (much as affluent people in domestic societies have special responsibilities, such as to pay higher tax rates to fund social welfare programmes) – but precisely which states have precisely how much responsibility has never been agreed, partly because some of them look to newly developed states (China being the most obvious example) as being a bigger part of the problem and being more capable of taking on greater responsibility.

Nationally, governments have started to take concerted action on climate change. Policies have been put in place in most developed states to reduce (or at least slow) greenhouse gas emissions. This has been done through a range of policies, such as regulation of energy-generation and transport industries, taxes, subsidies and education. Even in most developed states where there is scepticism among some politicians about taking action to reduce greenhouse gas emissions, most notably the United States (wherein, under President Donald Trump, the federal government has actively sought to undermine existing climate-related policies, prevent new ones and even promote the extraction, use and export of fossil fuels), greenhouse gas emissions have started to stabilize or decline. Partly this is due to the actions of sub-state actors, such as municipalities, as well as efforts by some industries either to become more energy efficient or to capitalize on alternative sources of energy. Many developed states have also provided funding – in keeping with their common but differentiated responsibilities – to developing states to help them prepare for climate change and develop in more environmentally sustainable ways, although the levels of funding for these purposes has been vastly less than what is needed (see Harris 2016b: 172–6).

Alas, despite these and countless other policies and actions to mitigate greenhouse gas pollution, global emissions are still growing (Pierre-Louis 2018). Indeed, even if all of the world's national governments were to fully implement their Paris Agreement pledges to limit those emissions, the objectives of that agreement – to keep global warming to no more than 2°C above historical levels, and to try to limit warming to 1.5°C – would not be achieved; far from it (Climate Action Tracker 2018). Global warming has already exceeded 1°C (IPCC 2018: 51), and current policies will likely lead to global warming of well over 3°C (Climate Action Tracker 2018), to devastating effect. Even achieving the 1.5°C objective of the Paris Agreement would not prevent very major impacts and suffering around the world (IPCC 2018). In short, the problem is getting worse. Short of radical policy changes all around the world, the prospects are that the pollution causing climate change will continue for decades to come. Adding insult to injury, there is currently very little prospect of developed-state governments increasing financial and other assistance to developing states enough for the latter to cope adequately with the impacts of climate change. Without dramatic changes to the ways in which societies function, especially with respect to energy use, the injustices of climate change will be exacerbated as time passes.

Climate justice

Justice is about fairness, equity, impartiality and doing what is morally right. If something is unfair, inequitable, immoral or unreasonably partial (especially against those who are weak or toward those who are powerful or otherwise advantaged), it may be deemed to be unjust. With this in mind, in this book we conceive of *climate* justice broadly in terms of the fairness, equity and rightness of responses to climate change. (We do not focus on legal justice per se, although legal issues, both national and international, are clearly germane and thus do feature to varying degrees in many of the chapters.) We can perhaps understand the place of justice (and injustice) in the context of climate change most easily by thinking about relationships among actors, such as individuals, groups, organizations, corporations and government agencies, in local communities. If, within our communities, those with wealth and power are exploiting those with neither, we would normally describe that exploitation as unjust (despite the reality that it may occur as a matter of routine). More to the point, if a rich individual within a community were to pollute environmental resources relied upon by everyone in the community, for example by dumping pollution into a river that supplies drinking water to everyone downstream, most people would say that such activity is unjust (and many courts would agree because such activity would likely be unlawful). If a rich city were to dump its toxic rubbish onto a poor community, we might call that unjust, too. And if the hardship that arises from such activity were not compensated (and perhaps even if the perpetrators were not punished), most people would be inclined to say that justice has been denied.

All of these sorts of injustices, and many more, are occurring in the context of climate change, both within communities and globally. For example, climate change is a matter of justice because it involves the distribution of benefits and burdens globally and nationally. Some states and people are enjoying the benefits of polluting the global atmosphere while the burdens of doing so are being borne disproportionately by other states and peoples. Climate change is a matter of justice because its human causes will result in impacts that bring loss, hardship and suffering to many millions – and perhaps ultimately to billions – of people. Climate change is a matter of justice because the people who are least responsible for causing it, especially the world's poor, will suffer the most from it. Climate change is a matter of justice because the world's richest states and its most affluent people are disproportionately responsible for causing it. Climate change is a matter of justice because those who cause it often do so for non-essential, even frivolous reasons. Climate change is a matter of justice because those that do the most to cause it have the greatest capabilities to greatly reduce or stop their causative behaviours. Climate change is a matter of justice because those who are causing it are not doing nearly enough to aid those who do and will suffer from it. Climate change is a matter of justice because people in the present are voluntarily or unnecessarily harming people in the future. Climate change is a matter of justice because many people are behaving in ways that do, and increasingly will, undermine and destroy natural habitats and lead to the extinction of countless other species. Climate change is a matter of justice for countless other reasons.

A blossoming literature on climate justice

This book is intended to build upon existing literature on climate justice, particularly that which has appeared since the turn of the century. In recent years, the number of publications on climate justice has ballooned. This has been manifested in hundreds of scholarly journal articles on every aspect of climate justice (for a curated selection of several dozen articles within this genre, see Harris 2016a). From being a relatively rare topic in book-length literature on climate change in the 1990s, it is now commonplace for climate justice to be given at least some attention in most books about climate change. Quite a number of books published since the turn of the century are dedicated largely or wholly to climate justice. These include, for example, Page (2006), Vanderheiden (2008), Harris (2010), Humphreys (2010), Posner and Weisback (2010), McKinnon (2011), Shue (2014) and Tremmel and Robinson (2014).

Examples of books from just the last five years (2015–19) that focus much or all of their attention on climate justice include, among others, Aaron Maltais and Catriona McKinnon's (2015) edited book, which devotes several chapters to questions of climate equity and justice; Jeremy Moss's (2015) analysis, in collaboration with other scholars, of key problems associated with climate justice; Wen Stephenson's (2015) description of the lives of activists trying to achieve climate justice; Dominic Roser's (2016) introduction to climate justice and his co-edited volume with Clare Heyward (Heyward and Roser 2016) that considers the role of climate justice, and related questions of moral theory, in non-ideal circumstances; Randall S. Abate's (2016) collection of case studies on the regional and global implications of the disproportionate burdens of climate change that are felt by poor and marginalized people around the world; Christopher J. Preston's (2016) edited volume on the implications for climate justice of geoengineering (i.e., intentional large-scale efforts to 'engineer' Earth's climate system); Paul G. Harris's (2016b) survey and critical analysis of cosmopolitan 'global' justice and climate change (see his chapter with Alix Dietzel in this book); Luke Tomlinson's (2016) detailed study of procedural fairness in negotiations leading to the UNFCCC; Tracey Skillington's (2017) look at how denial of human rights has resulted in climate change and how strengthening human rights might promote climate justice; Olivier Godard's (2017) exploration of the 'diversified and conflicting intellectual landscape of the idea of global climate justice' (Godard 2017: 4); and Lukas H. Meyer and Pranay Sanklecha's (2017) edited volume in which scholars analyse the central debates regarding responsibility for historical emissions of greenhouse gases and what this means for the formulation of just climate policies.

The number of books on climate justice appearing in the last year or so (2018–19), not to mention those that are forthcoming, is remarkable. They include, for example, Mary Robinson's (2018) memoir, in which she argues that addressing climate change is essential to realizing global justice; Byron Williston's (2018) exploration of international and intergenerational justice as central themes in the ethics of climate change; Anna Roosvall and Matthew Tegelberg's (2018) investigation of the nexus of media and indigenous peoples in 'transnational' climate justice; Stefan Gaarsmand Jacobsen's (2018) collection in which a number of scholars look at the economic aspirations of the global movement for climate justice; R. Scott Frey, Paul K. Gellert and Harry F. Dahms's (2018) edited book in which several chapters look at some of the injustices of climate change, especially as they relate to historical global inequalities; Tahseen Jafry's (2018) comprehensive edited 'handbook' dedicated to climate justice, which surveys much of the landscape of this topic; Alix Dietzel's (2019) cosmopolitan-oriented look at the roles that non-state actors play in climate action and justice (see her chapter with Paul G. Harris later in this book); John S. Dryzek and Jonathan Pickering's (2019) book on the politics of the Anthropocene – the current period in Earth's history, during which humans have become the decisive force in shaping the Earth system, particularly its ecology – in which the authors focus substantial attention on what they call 'planetary justice'; Ravi Kanbur and Henry Shue's (2019) collection of chapters that brings together the perspectives of economists and philosophers on how to understand and implement climate justice (Shue is arguably the most prominent scholar of climate justice, with much of his working building on his argument for 'basic rights' from the 1980s (see Shue 1996, 2014)); and Megan Blomfield's (2019) development of a theory of global resource egalitarianism that, like Dietzel (2019) and Harris's (2016) books, challenges traditional conceptions of state sovereignty in the context of climate change.

The upshot is that climate justice research has joined mainstream scholarship on climate change. This is a positive development, but the knowledge and awareness that have been created in this area remain inadequate. Research and writing about climate justice are on the increase, but so, too, are the injustices of climate change. Much more research is needed, and soon, and vastly more effort is needed to translate resulting ideas into action. The chapters in this book signify steps in that direction.

Agendas for climate justice

In the chapters that follow, the contributors present innovative ways of conceptualizing and realizing climate justice. They critique mainstream ways of thinking about (and theorizing) climate justice. They attempt to push the boundaries of prevailing practice on climate justice. Many of their arguments are intentionally provocative – because provocation may be what is needed to shake both scholars and practitioners out of established ways of thinking and customary ways of acting that have, for decades, failed to prevent most of the injustices of climate change from increasing.

Vital needs of humans and non-humans

We begin in Chapter 2 with an exploration of vital needs and climate justice by Robin Attfield. Attfield argues that we need to consider inter-human, intergenerational and inter-species perspectives when building the foundations for climate justice. Systematic studies of environmental (in)justice, beginning decades ago, have revealed that environmentally harmful activities are often located close to poor and dispossessed people. Such environmental injustices are now widespread, an obvious example being the placement of garbage dumps in the vicinity of (or even within) poor communities (and never within the richest ones). Such activities engender substantive, procedural and recognitional injustices. The growing environmental justice movement, in addition to emphasizing the impacts of environmental pollution on local communities, has also concerned itself with justice for future generations. According to Attfield, despite problems associated with explicating the rights of those living in the future, there might be circumstances in which they can justifiably be treated unfairly, as may occur when the interests – particularly vital ones – of those living in the present are prioritized while the interests of future generations are relatively discounted. From this perspective, preventing harm to future generations could amount to its own kind of injustice. Understanding why this could be the case, and getting a full picture of the scope of climate justice, requires us to consider the vital needs of other species alongside those of humans. To omit non-human creatures from the scope of justice could result in unjustly sacrificing the *vital* needs of non-human creatures to *peripheral* interests of humans. In evaluating this sort of perspective, Attfield maintains that environmental ethicists and policy-makers, if they are genuinely concerned about climate justice, ought to take all of these interests – human *and* non-human – fully into account.

Identifying responsible agents

In Chapter 3, Ivo Wallimann-Helmer critically examines agency for climate change in the context of common but differentiated responsibilities. He points out that ethical challenges presented by climate change routinely involve the just distribution of benefits (entitlements) and burdens, on one hand, and the fair differentiation of responsibilities, on another. The fairness of the latter will rely on principles of justice. The applicability of those principles, and the demands that they make, are dependent on several factors: the policy domains for mitigation, adaptation and 'loss and damage' associated with climate change; the agents (routinely identified as states) that bear the responsibilities; and the relevant policy levels – international, regional, national and local – of climate action. Wallimann-Helmer argues that the principle of common but differentiated responsibilities is more than a starting point for climate justice. It also shapes which demands for action on climate change are most appropriate for the different agents responsible for such action.

Worldviews and climate justice

From the perspective of David E. Storey, climate justice is and ought to be overwhelmingly cosmopolitan. (In Chapter 7, Alix Dietzel and Paul G. Harris advocate a cosmopolitan agenda for climate justice.) That said, in Chapter 4 Storey argues that the failure of ethics to penetrate sufficiently into climate policy, and indeed the inability of comprehensive and effective climate policy to gain political traction, requires something more. He argues that we ought to stop asking whether our accounts of climate justice are right or wrong and instead start asking why they have not been more effective. His chapter lays out a research agenda that consolidates moral philosophy and the social sciences to develop a potentially more effective approach to climate justice. He does this by first identifying three worldviews – traditional, modern and postmodern – that arguably have fuelled political and policy gridlock on climate change. He shows that debates about climate justice have often been framed as zero-sum conflicts between people who have 'modern' and 'postmodern' worldviews, on one side, and what he believes to be largely ignored people who have 'traditionalist' worldviews, on the other. To move beyond this conflict, and thereby to realize climate justice, Storey believes that it is necessary to find ways of framing justice and climate change, and also of communicating the benefits of climate policy, in ways that are likely to bring about a congruence of perspectives among the major worldviews.

Spatial framing of injustices

In Chapter 5, Ian Bailey also identifies different (and conflicting) ways of perceiving climate change as obstacles to realizing climate justice. He examines how notions of (in)justice are contested in national climate politics and looks at how competing conceptualizations of justice have shaped the development of policies intended to bring about decarbonization of national economies. He focuses on the role of 'spatial anchors' in the legitimization of justice arguments for (or against) new policy measures to address climate change. He does this by drawing upon evidence from actual experience in Australia, New Zealand, the United Kingdom and the United States. Analysis of that evidence reveals to Bailey that advocates of stronger mitigation policy frequently emphasize broader-scale concerns about the responsibility of wealthier states to take action. In contrast, those who want to obstruct or dilute climate policy initiatives often stress national welfare, inaction by other states or local justice concerns. The effectiveness of those who oppose robust climate policies in constructing spatially and socially recognizable discourses about the injustices of climate action has created major obstacles to climate-protecting policies. It has also undermined the influence of climate justice on political agendas. To increase the likelihood of realizing climate justice, Bailey argues that greater attention should be given to representing justice arguments in spatially imaginative ways.

Capitalism, degrowth and climate justice

In Chapter 6, Carlos Tornel also critiques prevailing ways of conceiving of climate justice and their failure to restrain the global forces, especially material economic growth, that are causing and exacerbating climate change. Indeed, Tornel argues that environmental non-governmental organizations, and the mainstream conceptions of climate justice that are advocated by an elite international managerial class, have relatively little beneficial significance. As he sees it, the dominant approaches to climate justice have systematically depoliticized, and thereby neutered, the demands of the climate justice movement. Climate change has been conceived of as a problem that requires technocratic and managerial solutions. This conceptualization has perpetuated a 'post-political' condition in which elite actors become involved but 'nothing really changes'. In response, in this century the objective of ending material economic growth – degrowth – has emerged as a social and academic movement. According to Tornel, a repoliticization of the climate justice movement is necessary to achieve climate justice, and degrowth offers a way to bring that about. The degrowth movement can do this by critiquing prevailing techno-managerial discourse, radically contesting calls for individual responsibility and collective action, and demanding revolutionary social transformation. For Tornel, solving the climate crisis and realizing climate justice requires moving beyond highbrow technological solutions, which most often arise from dominant discourses of ecological modernization and capitalism, toward genuine degrowth throughout the global economy.

Cosmopolitan justice and non-state actors

In Chapter 7, Alix Dietzel and Paul G. Harris explore the role of individuals and other non-state actors in doing the things that are necessary to realize climate justice. They argue that looking at climate change from a cosmopolitan perspective can help us to fully comprehend the practical and normative complexities of the problem. A cosmopolitan viewpoint discerns these actors' responsibilities in efforts to mitigate climate change, adapt to its impacts and aid those who do and will suffer from it. An agenda for cosmopolitan justice would help to move research and action beyond considering what states can and should do about climate change to focus much more attention on what non-state actors can and should do, and to some extent already are doing. As Dietzel and Harris note, there has been an explosion in the number of non-state actors involved with climate change. What is more, billions of individuals are capable of doing far more than they are at present. The climate regime, particularly following the 2015 Paris Agreement, has started to treat non-state actors as key agents of effective action. A cosmopolitan agenda for climate justice critiques states as exclusive holders of climate-related responsibility and vehicles for climate justice. Individuals and other non-state actors are also responsible and able to contribute to realizing climate justice. According to Dietzel and Harris, research for climate justice ought to take this fully into account.

Social justice and ecological consciousness

For individuals to take the action that is necessary to realize climate justice, it is necessary for them to be more ecologically conscious. As James S. Mastaler reminds us in Chapter 8, there are many profound moral and social consequences arising from humanity's collective failure to reduce global greenhouse gas emissions. This failure has resulted in fundamental injustices against the world's poorest communities, which Mastaler argues necessitates new kinds of responses that are grounded on social equity and environmental responsibility. Together, these social and environmental attributes can form the foundation for the emergence of much greater ecological consciousness. Mastaler contemplates the development of ecological consciousness among individuals and across institutions, and he proposes a programme of essential tasks for realizing climate justice sooner rather than later. He introduces his own long-term vision for furthering climate justice. Social justice and ecological consciousness will be features of the structural changes that will have to be implemented if this is to happen.

Democratic institutions and environmental citizenship

In Chapter 9, Giuseppe Pellegrini-Masini, Fausto Corvino and Alberto Pirni also explore institutional changes, focusing their arguments on the realization of climate justice through environmental citizenship. They believe that two key problems prevent individual and uncoordinated actions from addressing climate change effectively. The first problem is that epistemological and moral complexities often prevent individuals from fully comprehending their responsibilities toward present and future generations. The second problem is that even those individuals who overcome the first problem may fail to act – that is, change their behaviours to reduce their own greenhouse gas emissions – because they are caught up in habits of procrastination that are intended to avoid the relatively high short-term costs that they would face in doing so. Pellegrini-Masini, Corvino and Pirni introduce a number of institutional mechanisms and policies for fostering climate citizenship that could be implemented to help people overcome these problems. Climate-citizenship policies, such as personal carbon allowances, ombudspersons for future generations and parliamentary representational quotas for them, have the potential to influence citizens' climate-related behaviours, thereby assisting in the realization of climate justice.

Justice for climate refugees

In Chapter 10, Justin Donhauser examines one of the real-world manifestations of climate injustice: climate refugees. He asks whether it is necessary to develop new institutions and frameworks for recognizing the rights of climate refugees and the duties of those that ought to help them. He concludes that new mechanisms are not necessarily required. Avenues for realizing effective and just responses to increasing numbers of climate refugees can be found in existing United Nations refugee and climate-policy mechanisms, suggesting that action on this growing problem can and should occur soon. Donhauser looks at many of the relevant normative justice concerns that are unique to cases of climate refugees. He lays out typical legal explanations for not extending to climate refugees the rights to asylum, relief and non-refoulement (i.e., the right not to be sent back to a place from which one is fleeing persecution) that are currently available to other kinds of refugees. His chapter assesses five proposals for addressing climate refugees and proposes some ways by which those proposals might be put into practice to realize justice for them. Donhauser advocates addressing these issues using existing provisions of the UN Refugee Convention and the UNFCCC. He supports his argument by using new climate-event modelling methods that prioritize which matters of climate justice are the most urgent.

Humanitarian crime and pre-emptive justice

In Chapter 11, Selina Rose O'Doherty argues that climate change is a humanitarian crime. She highlights the correlation between climate justice in moral terms and climate justice in more strictly legal terms. As O'Doherty notes, international agreements to address climate change identify future targets for action rather than taking immediate steps to mitigate climate-changing pollution. This is apparently justified by some actions because the impacts of climate change being felt currently were caused by past pollution, and because the climatic impacts of behaviours occurring today will not be felt anytime soon. O'Doherty argues that this relatively slow and future-oriented approach to climate change is unjust. She asks us to imagine what the appropriate response would be if the consequences of climate change were caused deliberately and with contemporaneous effect. Her answer: those consequences would qualify as breaches of global justice. From this perspective – viewing the actions that cause climate change as intentionally violating the human rights of millions of people – would mean that they are crimes against humanity. Thus, they can be classified as humanitarian climate crimes. O'Doherty makes a case for pursuing climate justice on behalf of future victims through pre-emptive humanitarian intervention against actors that are responsible for causing climate change. International intervention against humanitarian climate crimes would be justified as both an ethical obligation and a sovereign responsibility under currently accepted international standards of humanitarian protection.

Equity and justice in national pledges

Most of the chapters in this book undertake broadly theoretical arguments and qualitative analyses to arrive at new ways of approaching climate justice. In contrast, in Chapter 12, Claire Swingle performs an empirical content analysis of Nationally Determined Contributions (NDCs) – that is, national action pledges – of the Paris Agreement on climate change to demonstrate nominal national attitudes toward questions of climate equity. To better understand how national governments conceptualize climate justice, Swingle analyses the 'Fairness and Ambition' sections of 163 NDCs. She compares each state's indicators of equity in its NDC against historical positions of its respective negotiating group in the international negotiations on climate change. She further compares the NDCs to each state's submission under the Talanoa Dialogue (the facilitative dialogues that were convened in 2018 to take stock of states' actions on climate change and inform the preparation of updated NDCs in 2020). Swingle's comparative analysis reveals some surprising findings. For example, historical responsibility, the mantra of developing countries for decades, is underemphasized in their NDCs and largely disconnected from common but differentiated responsibility. Instead, objectives to keep global temperatures below certain thresholds, and references to the importance of scientific reports, have been used in NDCs to build the case for enhanced ambition to reduce greenhouse gas pollution and to provide international support for developing countries. These findings are significant because the ways in which national governments perceive equity will affect how they assess the success (or lack of it) of global action on climate change, and subsequently how they shape their national climate policies. Those actions and policies will be crucial to realizing climate justice.

Responsibilities of the great powers

In Chapter 13, Sanna Kopra argues that the great powers – China, the United States and other states recognized as having especially powerful status in international affairs – have greater responsibility for promoting climate justice than do most other states. He does this by drawing on the English School of International Relations theory, which posits that states form an international society and that great powers have special responsibilities in that society. By Kopra's reading, the English School has not paid very much attention to climate justice despite the fact that climate change is creating risks for international security and justice, and indeed will do so much more in the future. This is a gap in the literature that Kopra aims to fill in part with his chapter. He proposes that great powers can be expected to shoulder two types of special responsibility for climate change: managerial responsibility to prevent climate-related conflicts from undermining international order, and leadership responsibility to promote human values through ambitious global climate policies. However, as Kopra points out, China, the United States and the UN Security Council (made up of great powers) have all failed to define their great-power responsibilities ambitiously enough to prevent dangerous climate change. Kopra advocates urgent research to develop the English School's normative viewpoints of great power responsibility and to use those to inform international negotiations on climate change and associated national policies. Realizing climate justice will require this.

Conclusion

The justice implications of climate change have been widely debated for decades. Calls for justice, particularly international justice, have been written into climate change agreements and conventions. Efforts to limit the injustices of climate change and to highlight them are also now part and parcel of the work of many non-governmental organizations, especially those concerned with environmental protection, economic development and poverty eradication. Increasingly, even some businesses have accepted that climate justice demands certain things of their behaviours. Appropriately, climate justice is now an established area of scholarship that crosses disciplinary boundaries, drawing the attention of natural and social scientists alike. Yet, despite the work of governments, scholars and activists to study and implement climate justice, the injustices of climate change – manifested in continued greenhouse gas pollution and, increasingly, the felt impacts of the changes they bring about – continue to worsen. Short of sweeping changes in individual and collective behaviours (which arguably would be worthwhile in any case because they can improve human well-being in numerous ways), there is little prospect of averting very serious climate change in the future.

Climate change is not only an environmental problem, nor is it primarily a problem of economics or politics. Climate change is fundamentally a problem of justice: injustice is at the root of its causes, at the heart of its impacts and vital to whether and how effective policies will be devised and implemented to mitigate the associated hardships. It would not be far-fetched to say that climate change is rapidly becoming the greatest injustice ever witnessed, experienced and indeed perpetrated across all of human history. Realizing climate justice under current circumstances will require doing far more in the very near future, and it will necessitate sustained action for many decades to come (at least). It will require new and innovative vision about the way forward. Without a new agenda – or new agendas – for researching, and therefore understanding, climate justice, not to mention accompanying new ways of acting to realize climate justice, the injustices of climate change will only multiply and intensify.

Each of the chapters that follow advocates its own potential research agenda for climate justice. Looked at collectively and holistically, they present a single research agenda for climate justice that is characterized by more attention (theoretically and practically) to justice in the context of climate change and, more specifically, greater effort to arrive at and implement alternative, even radical, conceptions of climate justice, sooner rather than later. Justice demands nothing less.

References

Abate, Randall S. (2016), *Climate Justice: Case Studies in Global and Regional Governance Challenges*, Washington, DC: Environmental Law Institute.

Blomfield, Megan (2019), *Global Justice, Natural Resources and Climate Change*, Oxford: Oxford University Press.

Climate Action Tracker (2018), 'Warming projections global update', accessed 29 May 2019 at https://climate actiontracker.org/documents/507/CAT_2018-12-11_Briefing_WarmingProjectionsGlobalUpdate_ Dec2018.pdf.

Dietzel, Alix (2019), *Climate Justice and Climate Governance: Bridging Theory and Practice*, Edinburgh: Edinburgh University Press.

Dryzek, John S. and Jonathan Pickering (2019), *The Politics of the Anthropocene*, Oxford: Oxford University Press.

Frey, R. Scott, Paul K. Gellert and Harry F. Dahms (eds) (2018), *Ecologically Unequal Exchange: Environmental Injustice in Comparative and Historical Perspective*, Cham: Palgrave Macmillan.

Godard, Olivier (2017), *Global Climate Justice: Proposals, Arguments and Justification*, Cheltenham, UK and Northampton, MA, USA: Edward Elgar Publishing.

Harris, Paul G. (2001), *International Equity and Global Environmental Politics*, London: Routledge.

Harris, Paul G. (2010), *World Ethics and Climate Change*, Edinburgh: Edinburgh University Press.

Harris, Paul G. (ed.) (2016a), *Ethics, Environmental Justice and Climate Change*, Cheltenham, UK and Northampton, MA, USA: Edward Elgar Publishing.

Harris, Paul G. (2016b), *Global Ethics and Climate Change*, 2nd edn, Edinburgh: Edinburgh University Press.

Harris, Paul G. (2018), 'Climate change: science, international cooperation and global environmental politics', in Gabriella Kutting and Kyle Herman (eds), *Global Environmental Politics: Concepts, Theories and Case Studies*, 2nd edn, London: Routledge, pp. 123–42.

Harris, Paul G. (ed.) (2019), *Climate Change and Ocean Governance: Politics and Policy for Threatened Seas*, Cambridge: Cambridge University Press.

Heyward, Clare and Dominic Roser (eds) (2016), *Climate Justice in a Non-Ideal World*, Oxford: Oxford University Press.

Humphreys, Steven (ed.) (2010), *Human Rights and Climate Change*, Cambridge: Cambridge University Press.

Intergovernmental Panel on Climate Change (IPCC) (2013), *Climate Change 2014: Synthesis Report*, Cambridge: Cambridge University Press.

Intergovernmental Panel on Climate Change (IPCC) (2018), *Global Warming of 1.5 °C*, accessed 29 May 2019 at https://www.ipcc.ch/sr15/.

Jacobsen, Stefan Gaarsmand (2018), *Climate Justice and the Economy: Social Mobilization, Knowledge and the Political*, Abingdon: Routledge Earthscan.

Jafry, Tahseen (ed.) (2018), *Routledge Handbook of Climate Justice*, London: Routledge.

Kanbur, Ravi and Henry Shue (eds) (2019), *Climate Justice: Integrating Economics and Philosophy*, Oxford: Oxford University Press.

Maltais, Aaron and Catriona McKinnon (eds) (2015), *The Ethics of Climate Change*, London: Rowman and Littlefield.

McKinnon, Catriona (2011), *Climate Change and Future Justice: Precaution, Compensation and Triage*, London: Routledge.

Meyer, Lukas H. and Pranay Sanklecha (eds) (2017), *Climate Justice and Historical Emissions*, Cambridge: Cambridge University Press.

Moss, Jeremy (2015), *Climate Change and Justice*, Cambridge: Cambridge University Press.

Page, Edward A. (2006), *Climate Change, Justice and Future Generations*, Cheltenham, UK and Northampton, MA, USA: Edward Elgar Publishing.

Pierre-Louis, Kendra (2018), 'Greenhouse gas emissions accelerate like a "speeding freight train" in 2018', *New York Times*, accessed 24 May 2019 at https://www.nytimes.com/2018/12/05/climate/green house-gas-emissions-2018.html.

Posner, Eric A. and David Weisback (2010), *Climate Change Justice*, Princeton, NJ: Princeton University Press.

Preston, Christopher J. (ed.) (2016), *Climate Justice and Geoengineering: Ethics and Policy in the Atmospheric Anthropocene*, London: Rowman and Littlefield.

Robinson, Mary (2018), *Climate Justice: Hope, Resilience and the Fight for a Sustainable Future*, New York: Bloomsbury.

Roosvall, Anna and Matthew Tegelberg (2018), *Media and Transnational Climate Justice: Indigenous Activism and Climate Politics*, New York: Peter Lang.

Roser, Dominic (2016), *Climate Justice: An Introduction*, Abingdon: Routledge.

Shue, Henry (1996), *Basic Rights*, 2nd edn, Princeton, NJ: Princeton University Press.

Shue, Henry (2014), *Climate Justice: Vulnerability and Protection*, Oxford: Oxford University Press.

Skillington, Tracey (2017), *Climate Justice and Human Rights*, New York: Palgrave Macmillan.

Stephenson, Wen (2015), *What We're Fighting for Now is Each Other: Dispatches from the Front Lines of Climate Justice*, Boston, MA: Beacon Press.

Tomlinson, Luke (2016), *Procedural Justice in the United Nations Framework Convention on Climate Change: Negotiating Fairness*, New York: Springer.

Tremmel, Joerg Chet and Katherine Robinson (2014), *Climate Ethics: Environmental Justice and Climate Change*, New York: I.B. Tauris.

United Nations Framework Convention on Climate Change (UNFCCC) (1992), 'United Nations Framework Convention on Climate Change', New York: United Nations, accessed 31 May 2019 at http://unfccc.int/files/essential_background/background_publications_htmlpdf/application/pdf/conveng.pdf.

Vanderheiden, Steve (2008), *Atmospheric Justice: A Political Theory of Climate Change*, Oxford: Oxford University Press.

Williston, Byron (2018), *The Ethics of Climate Change: An Introduction*, London: Routledge.

2 Vital needs and climate change: inter-human, inter-generational and inter-species justice

Robin Attfield

Research on climate justice needs to find ways to uphold the climate-related needs of present and future generations of human beings and of the other species with which we share the planet. In this chapter I explore the relevance of the vital needs of human beings and of non-human creatures (both present and future) to climate justice, beginning with conceptual links between justice and vital needs. I proceed to elicit these needs and give some examples of how they can be upheld.

Justice and vital needs

Central to justice is the meeting or satisfaction of needs. Needs are whatever is necessary to continued life (survival) and to well-being. I am not using 'needs' to mean fashionable clothing or the accessories of high culture, despite widespread tendencies (often originated by advertisers) to treat such items as 'needs'. Well-being on any account involves the development of and ability to exercise the characteristic capacities of one's kind. Hence, needs include (for humans) not only food, clothing and shelter, but also companionship and education. *Vital* needs are those needs that are indispensable, either for survival or for well-being. While some needs are needed to enhance well-being, such as the need of some musicians for an electric guitar, which is needed if a particular kind of musical capability is to be developed, there are others (vital needs) in the absence of which (human) well-being is at risk, such as the need for participation in one or another culture. Some of these vital needs are constitutive of well-being, such as the ability to participate in dialogue (of one kind of another); others are instrumental, making the satisfaction of other key capacities possible. For most human beings, one of these instrumental needs is access to a reliable electricity supply (discussed below).

While needs are central to justice, justice has other requirements or presuppositions. Thus actions, policies and practices can only be assessed as just or unjust where someone (singular or plural) or some organization is responsible for their actions and choices and able to generate, reject or modify the actions, policies and/or practices in question. The nature and role of responsibility are subject to multiple refinements, but these have to be set aside for current purposes; what is crucial is that praise for justice or blame for injustice presupposes responsibility.

Justice is concerned not only with needs but also with desert. The nature of desert is another matter eligible for extended debate. One defensible view is that desert is itself meritorious only if those embodying desert are thereby contributing to the satisfaction of needs. If this view is justified, then the centrality of needs is further underlined. But it is unnecessary to resolve this debate for the centrality of the satisfaction of needs to emerge, even if the satisfaction of needs is not what makes desert deserving or meritorious.

The centrality of vital needs is related to climate justice. Even if vital needs are restricted to those of contemporary human beings, certain policies turn out to be vitally needed, and thus to be requirements of justice. The needs of the people of developing countries cannot be forgotten if justice is to be done, and the required policies must take them into account. But the foreseeable needs of future human generations are equally important, and if they are to be satisfied, the policies must be sustainable ones. However, the needs of members of other species are arguably as central to justice as those of human beings, and the foreseeable needs of their future generations cannot be forgotten. Just policies will be moulded by all these needs, and by vital ones in particular. The contours of such policies are revisited towards the end of this chapter.

Climate justice: mitigation and adaptation

The Environmental Justice Movement has shown how environmentally harmful activities are often located close to the homes of poor and dispossessed people, both within the USA and Europe and also on or close to the coasts of Asia and Africa, where toxic waste has often been dumped. James Sterba has reasonably called these practices 'environmental racism' (Sterba 1998; see also Kelbessa 2012, and First National People of Color Environmental Leadership Summit 1991). These are discriminatory practices, and they are usually conducted without the consent of those affected; thus, they raise issues of substantive justice and of procedural justice as well. Besides, even when there are procedures for consultation, effectively to ignore powerless minorities and people despite the availability of such procedures further raises issues of recognition (or the lack of it), and thus of recognitional justice.

Future people, of course, cannot be consulted or recognized (except in the sense of young people already alive). But even unborn future people can be treated unfairly, when they are ignored in policy decisions affecting their quality of life. This is not strictly a matter of heeding their rights, for reasons explained by Derek Parfit (1984), for their very existence may depend on policies adopted in the present, and so they cannot strictly be harmed, nor their rights infringed. (See Parfit's example of Depletion in Parfit 1984, and O'Neill et al. 2018.) But many current policies can adversely affect the quality of life of whoever lives in future decades and centuries, and in matters of vital needs (and other needs) at that. Hence, decisions and policies that affect the realization of the needs of future generations can, for these reasons, be unjust, something which the Environmental Justice Movement is free to recognize.

Yet policy decisions about mitigating carbon dioxide and other greenhouse gas emissions, and about adaptation to levels of those emissions that are by now unavoidable, affect both the whole generation of our contemporaries, and the realization of their needs, whether they are responsible for those emissions or not, and also the life-prospects, and thus the vital needs, of our successors for many generations to come. Hence the issues of greenhouse gas mitigation and adaptation to emissions that are by now irreversible figure prominently within the fields of both international justice and inter-generational justice, and they are thus central to climate justice.

North–South relations and the vital needs of future generations

Climate justice involves not only mitigation of greenhouse gas emissions and adaptation to emissions that are either irreversible or unpreventable, but also sustainable provision for the needs of people in developed and developing countries across the coming generations. This is partly because of the necessary connection between justice and needs, and vital needs in particular, and partly because nothing short of sustainable provision for the satisfaction of needs can respond adequately to the needs of coming generations as well as of those of our contemporaries. Yet, current international relations fall far short of such provision. For example, to specify one commodity that is vitally needed by people everywhere, fresh water, many countries and regions have inadequate access to this resource, and the prospects for coming generations are far from close to sustainability. Droughts of increasing length and severity (caused in part by climate change) combine with the overuse of aquifers to exacerbate this problem.

Another vital need for most people is an adequate and reliable electricity supply. Without it, hospitals cannot be run, and most people can neither work nor study after dark. But satisfying this need will mean the generation of yet more electricity, which might seem poised to make the problems of carbon emissions worse. It really would mean this if the additional electricity were to be generated from carbon sources. Consequently, it is important, on grounds of climate justice, that the additional electricity needed by developing countries be generated from renewable sources. (Here I include China among developing countries, despite the strong case against such inclusion. See Harris 2011.) And where this is not currently feasible, justice requires that developed countries which already have the relevant technology should make this technology available to the relevant developing countries – that is, to countries of Asia, Africa, Latin America and Oceania which lack the ability to develop this technology unaided.

To turn to issues more immediately related to climate, current generations of these and other developing countries are undergoing the impacts of climate change, including direct impacts such as raised sea-levels and consequent flooding, and indirect impacts such as an increased frequency and intensity of adverse climate events such as storms, wildfires, droughts and floods. The injustice of this is underlined when it is remembered that most of the inhabitants of these countries have contributed little or nothing to the emissions of greenhouse gases responsible for the climate change that causes them.

In these circumstances, climate justice requires a range of responses from developed countries, as well as some from developing ones, and from countries until recently classified as developing countries, such as China and Malaysia, which are well on the way to the status of being developed countries. All countries are obliged to contribute to the mitigation of carbon emissions and the emission of other greenhouse gases. But where the adverse impacts just mentioned are visited on developing countries that have not generated the climate change from which these impacts derive, the countries responsible need to facilitate adaptation, not least in cases where relevant developing countries lack the resources to afford it. And the grounds for this claim are not limited to beneficence and charity; they extend to the requirement to make reparations to people whose situation is due in part to the wealth generated by earlier generations in developed countries through their deployment of carbon-based industrial processes, the very processes which underlie climate change.

Policies of adaptation may involve in some cases the provision of sea-walls and of raised highways and railways, or better bridges, and in others of improved infrastructure including enhanced electricity distribution systems, in addition to improved forms of electricity generation from renewable sources (for if the additional electricity is generated from carbon-based sources, the overall problem of climate change will be increased beyond even its current serious nature). The transfer of relevant technology should be mentioned again as having a key role if adaptation in developing countries, including those well on the way to being developed ones, is to be sufficient in extent and to be delivered before it is too late.

At the same time, the needs (and in particular the vital needs) of future generations need to be taken into consideration. Thus if average temperatures rise by more than 1.5 degrees (Celsius) above pre-industrial levels, then these generations are likely to have to undergo far more frequent and more intense adverse climate events, and flooding so severe that, in the case of small islands, their territory shrinks or disappears altogether (see UNFCCC 2018). Simultaneously, coastal settlements, including numerous large coastal conurbations, will be under increasingly severe threat of becoming uninhabitable.

But to anticipate and prevent the onset of such problems, and the related non-satisfaction of the vital needs of millions of our successors, requires both sustainable forms of adaptation, and strong policies of greenhouse gas mitigation, such that the more severe threats attaching to climate change are prevented from coming about. (More will be said about mitigation and adaptation below.) Additionally, planning for a sustainable future in these circumstances may include, on the one hand, programmes for the relocation of coastal communities and, on the other hand, resettlement programmes for the populations of islands such as Fiji, Vanuatu and the Maldives, future generations of whose inhabitants may find their territories inundated. Such resettlement is a matter calling for international agreement, but climate justice, in view of the responsibilities specified above, requires developed countries to be willing to admit into their territories significant groups of people dispossessed by the impacts of climate change. Nor are these measures likely to prove sufficient; for example, the deltas of rivers such as the Ganges and Brahmaputra, of the Nile, and of the Mississippi are going to need reinforcement of river banks, together possibly with the abandonment of human settlements in other places.

However, the kinds of relocation, resettlement and reinforcement just mentioned are likely to involve large-scale economic changes, in which many people's vital needs are unlikely to be satisfied unless much greater provision is made to satisfy the medical, educational and occupational needs of the generations planned for. Countries that have reconfigured their own territory in the past, such as the Netherlands, may be in the strongest position to advise the affected developing nations on how best to introduce appropriate policies. However, as all coastal developed countries will be addressing many of the same issues, because of the problems affecting their own coastlines, there should be an increasing pool of expertise in these countries, and climate justice will increasingly require that this expertise be made available to those developing countries that both lack and need it. Relevant expertise includes not only that of engineers capable of diverting rivers and strengthening river banks and bridges, but also that of social scientists with expertise in the problems of societies undergoing major transitions, whose skills are likely to be needed every bit as much.

It can be objected that the pursuit of universal human development (if understood as depending on economic growth) could potentially undermine the biospheric processes on which development itself (and much else) depends. If, however, the aim is not growth but sustainable and equitable human well-being, coupled with a reduction in resource use (O'Neill et al. 2018), then the attainment of the kind of development linked to the delivery of climate justice is far from impossible. The United Nations' Sustainable Development Goals (United Nations 2015) can be interpreted in this way, instead of the widespread alternative interpretation where the underlying main aim is growth. A different objection should also be considered. Derek Parfit has argued that we cannot have obligations to most future individuals, since the identities of most future individuals are currently unknown and unknowable. This is because they would not exist at all unless particular policy choices are made in the present. It might be thought to follow that we can have no obligations, whether of justice or of benevolence, to future generations, and thus that intergenerational justice is confined to relations between the current generation and those future individuals whose identities can currently be known because they have already been conceived.

However, Parfit himself has replied to this reasoning. For while there are no obligations owed to individuals of the future who are currently unidentifiable, there can still be obligations with regard to sets of future people, such as the inhabitants of particular regions in particular periods, or (come to that) in all periods. If we are in a position to make a difference to the quality of life of whoever lives in the future, then the mere fact that we cannot currently identify who will live does nothing to show that we lack obligations to make the quality of life of whoever lives better than it might have been if we had made different policy choices. While most future individuals cannot be harmed, or have their rights infringed by actions or inaction of ours, we still have obligations to prevent foreseeable and avoidable deteriorations in their quality of life (Parfit 1984, Part IV).

Parfit's reply here appears convincing; we cannot evade obligations, including obligations of justice, with regard to future generations, on the misguided basis that obligations are invariably restricted to what is owed to identifiable individuals. In the context of climate justice, Parfit's reply suggests that we have obligations to mitigate greenhouse gas emissions that are known to be likely to cause the inundation of coasts and small islands, and to increase the frequency and intensity of extreme climate events such as storms, hurricanes, droughts, floods and wildfires. Such extreme climate events are already becoming more intense and also more frequent, but our obligation to prevent them is grounded not only in the harm that they generate to our contemporaries, but also in the probably greater (and ever increasing) deterioration of quality of life that they are set to generate for our successors.

Justice and non-human species: obligations of justice

As soon as issues of obligations with regard to non-human creatures are raised, a surprising conflict is prone to emerge. On the one hand it is widely and increasingly agreed that it is wrong to inflict suffering on these creatures, whether through action or through neglect of the animals in one's care, and wrong not to take steps to prevent such suffering (Singer 1979 [1993], Attfield 1995). When it is wrong not to do something, it is widely agreed that that course of action is obligatory. Hence, it is implicitly agreed that there are obligations with regard to non-human animals to which most human agents are subject. Further, those who accept Peter Singer's (1993) claim that like interests should be treated alike irrespective of the species of the creature affected must also agree that when the vital interests of non-human animals and more trivial human interests are at stake, we are obliged to uphold the former at the expense of the latter, as having a higher priority.

On the other hand, it is often held that non-human animals fall outside the scope of justice (Rawls 1971 [1999]: 118–23 and 442–48; Rawls 1993: 109; Barry 1999: 95;

Scanlon 1998: 97–8; Möllendorf 2002: 31–6), and that justice gives rise to obligations to (some or all) human beings only. Hence any obligations due to non-human animals are trumped by these obligations of justice, and accordingly they carry a lower priority than obligations of justice. Some might infer that obligations with regard to non-human animals warrant our attention only after all obligations of justice to human beings have been performed. Such deprioritized obligations would, if so, constitute obligations no more than nominally.

Adherents of this second view are prone to support it through adopting a contract theory of justice. Those favouring a Rawlsian theory of justice are concerned with the choices that rational choosers would make in an ideal choice situation (Barry 1978, 204–48), but their very model excludes non-humans, since few if any of these are rational choosers, and because rational choosers would seldom if ever adopt rules for society that took non-human animal interests seriously. Other theorists likewise attempt to discern 'what we owe one another' through reflection on choices for which reasonable defences can be presented (Scanlon 1998), thus underlining that the 'we' in question are human beings, and that only those capable of either supplying or heeding reasonable defences are likely to have their interests taken into account. Other adherents of this second view might adopt a Kantian basis, and hold that the basis of justice is respect for human beings, whose needs or interests will thus be the exclusive focus of justice (Möllendorf 2002, 31–6). Indeed, such people may well find no place for heeding the needs of non-human animals even in sectors of ethics lying outside justice. Doubtless there will be yet further bases on which the second view could be held, but enough has already been said to show how it could be motivated.

However, if the fundamental concern of ethics and of justice alike is needs (and vital needs in particular), then a different understanding of justice emerges. On a biocentric understanding of moral standing, the needs of all living creatures have to be prioritized. But because biocentrism is a minority stance, and since a much more conservative stance would suffice to undermine the second view, let us consider the stance of sentientists, who restrict moral standing to sentient creatures. For adherents of this stance, the needs of sentient creatures are to be taken into consideration, and where vital needs are at stake, these are the needs that must be prioritized, whether they belong to human or to non-human creatures (Singer 1979 [1993]). Consider now what a vital-needs-based theory of justice will say. Such a theory cannot exclude the vital needs of non-human animals from the scope of justice; it is likely to uphold strong obligations to respect these vital needs, and thus obligations of justice to prevent suffering, or at least to minimize it.

Hence, if the approach of this chapter, with its central emphasis on vital needs, is accepted, then the vital interests of animals will not be excluded from the objects of obligations of justice, nor will they be trumped by advocacy of giving priority in ethics to obligations of justice over obligations of other kinds (such as obligations of compassion). All of this will, of course, clash with the implications of contract theories, and with most versions of Kantian theories too (other than variants of

Kantianism such as Schopenhauer's, which expands respect for persons to respect for sentient beings (Schopenhauer 1851 [1974]). So, if vital needs are to be central to our understanding of justice, then we have no choice but to discard both contract theories of justice, and most forms of Kantian ethics at the same time. This conclusion may lead some simply to reject the centrality of vital needs in order to uphold either Kantianism or contract theory. However, the deeply arbitrary tendencies of this rejection and its implications for theories of justice may make many people disinclined to take such a heroic stance (see Attfield and Humphreys 2016: 1–11 and 2017: 44–57).

It is worth investigating what the implications are if the vital-needs-based understanding of justice is accepted. For it will follow that justice will now apply not only to inter-human relations among contemporaries, and not only to these relations and relations between humans of different generations, important as inter-generational justice continues to be. Justice will now apply in addition to relations between species, at least when vital needs are at stake.

The bearing of inter-species justice on climate change ethics

While the inter-generational nature of justice (and of morality in general) extends what used to be regarded as the scope and limits of morality, its inter-species nature extends it still more. This expansion of the scope of ethics was recognized in the last century by Hans Jonas (1984), who was also sympathetic to non-anthropocentric forms of normative ethics, such as those presented in the previous section of this chapter. Inter-species justice requires attention to regions of space that have seldom commanded much attention. Thus, the fish and other creatures living beneath the Antarctic ice-shield have an ongoing interest in that ice-shield not being eroded by human beings; and this turns out to be an issue not only of good ecological practice, but also of justice. Further, if there are living creatures on the moons of Jupiter or of Saturn, it emerges that we are required to take their interests into account, to the extent that we can affect them. But we should in any case pay heed to the interests of all of the creatures living in the forests of the Amazon, the Congo, Borneo and New Guinea, for example. In these cases, there is no doubt about their existence, nor about the threats to many of them from deforestation, from human-generated forest fires and from climate change. Indeed, the case for biodiversity preservation and that for action to mitigate climate change are markedly strengthened once inter-species justice receives recognition.

While inter-species justice makes some difference to the spatial scope of ethics, it makes much more difference to the corresponding temporal scope. For many current species are likely to survive the eventual extinction of humanity unless they are extinguished in coming decades; and so the interests of myriads of creatures, living both during the period of the human future and potentially after it, turn out not only to be subject to the impacts of human action and inaction in the present and the near future, but also to be relevant to the scope of ethics. If these species are eliminated by ourselves or by our near successors, then there will be no species-members in subsequent centuries, including those that could have lived after the demise of humanity, and this would mean that the flourishing that these creatures could have enjoyed will be prevented. The quality of life they would have had will be reduced to nil as a result of actions and policies performed or initiated by human action in the present.

Furthermore, those creatures which would have depended on these creatures for their sustenance are likely, even if they survive, to lead less vigorous and shorter lives, and thus to have a lower quality of life than they (or others of their generation) would have had, once again because of current decisions and actions. Consequently, the posthuman future needs to be included within the scope of ethics, and also within the scope of obligations of justice. In addition to this, the entire period of the future during which humanity survives falls within that scope as well. For, in addition to the quality of life of future people, to which (as we have seen) our climate policies are prone to make a difference, the quality of life of non-human creatures of this epoch also stands to be affected.

It has been estimated that there are currently some 8.7 million species, of which well below 2 million have been identified (Sweetlove 2011). Current climate policies are likely to have impacts on all of the future members of all of these species, both those that have been identified and those that have not. These organisms will all have vital needs, the fulfilment of which turns in part on current climate policies, and which appear to generate obligations of justice accordingly. However, since this claim presupposes some form of biocentrism, which many contest, the implications of sentientist theories of normative ethics should also be considered because they are more widely accepted. Sentientist theories imply that there are obligations of climate justice with respect to the quality of life of future sentient creatures, to which a considerable difference is likely to be made by current decisions and policies. While there are, and will in all probability be, far fewer sentient species in the future than living species of other kinds, the difference that current decisions and policies make to future members of those species remains considerable, as do the related obligations of climate justice.

Conclusion

Reasoning from the vital interests of distant and future human beings, and also from those of current and future non-human creatures, considerably strengthens the argument for policies that will stabilize greenhouse gas emissions, and limit average climate increases to 1.5 degrees (Celsius) above pre-industrial levels. At one stage there was agreement that the limit should be 2 degrees, but this agreement did not take into account the potentially terminal impact such an increase would have on small islands and their ecosystems, let alone the potentially terminal impact it would be likely to have on many of the vulnerable inhabitants of continents as well as of islands (UNFCCC 2018). The additional arguments specified above signify that the vital interests of future human beings and non-human creatures are to be added to those of small island communities in support of stronger policies of mitigation, the adoption of which is a requirement of climate justice.

Climate justice is not, however, restricted to mitigation. Much harm to our contemporaries and much deterioration of quality of life for future people and other creatures can be prevented by a range of adaptation policies, including assistance to developing countries which are not yet in a position to fund, plan or organize adaptation themselves. Nor is climate justice even restricted to mitigation and adaptation. For where past actions have undermined vital interests, the agents of those actions, and those who inherit the wealth and power acquired thereby, owe reparations to those affected and to their successors. This too is a requirement of justice, and, while it is normally regarded as owed to human beings only, can reasonably be understood as owed to non-human creatures too, in so far as creatures of the current generation are affected by past actions, such as destruction of habitat and greenhouse gas emissions, at any rate those of the period since 1990 when their impacts became widely understood and agreed.

The obligations of justice, where climate justice is concerned, thus turn out to be far-reaching, and cannot be set aside either on the basis that no issues of justice arise with regard to most future people, or on the basis that there are no obligations of justice with regard to non-human animals. These obligations considerably strengthen the case for policies of mitigation, of adaptation and of reparation, and at the same time the range of motivations available for supporting these policies.

Research into climate justice should further involve studying how best to preserve Earth's species (endangered ones in particular), and how to attain policies of sustainability which combine upholding current interests and those of successions of coming generations alike (cf. Attfield 2014, 2015, 2018). Such research has several dimensions. One concerns researching the most viable forms of sustainable agriculture, sustainable forestry, sustainable fisheries and sustainable cities. Another involves focusing on the quality of life of future human generations and on ways to prevent its deterioration. A third dimension of research focuses on studying the ecosystem-related and habitat-related needs of those wild species that scientists and others have identified, and the ways in which they might be satisfied despite their vulnerability to ongoing climate change. A fourth dimension is studying how to preserve those ecosystems in which hitherto unidentified species are most probably located, while a fifth dimension would aim to satisfy the vital needs of domestic animals whose prospects of well-being are largely dependent on their human custodians. A further form of research involves investigating to which human and non-human communities reparations are owed as a result of environmental and other forms of injustice in previous generations, and to what extent such reparations can be made to their current successors without generating new injustices to still other communities.

In addition to these forms of research, we are going to need to implement such research in the form of policies of sustainable agriculture, sustainable forestry, sustainable fisheries and sustainable community life, among other forms of sustainability, and also policies for stabilizing the human population itself, and maintaining it at a sustainable level. Research into all of these varieties of sustainability is one of the central emphases that research on climate justice is going to require, and the fruits of which policy-makers, governments and international organizations are going to need to pursue and promote.

References

Attfield, Robin (1995), *Value, Obligation and Meta-Ethics*, Amsterdam, the Netherlands and Atlanta, GA, USA: Editions Rodopi; since 2018 available from Leiden: Brill.

Attfield, Robin (2014), *Environmental Ethics: An Overview for the Twenty-First Century*, 2nd edn, Cambridge, UK and Malden, MA, USA: Polity Press.

Attfield, Robin (2015), *The Ethics of the Global Environment*, 2nd edn, Edinburgh: Edinburgh University Press.

Attfield, Robin (2018), *Environmental Ethics: A Very Short Introduction*, Oxford: Oxford University Press.

Attfield, Robin and Rebekah Humphreys (2016), 'Justice and non-human animals, Part I', *Bangladesh Journal of Bioethics*, 7(3), 1–11.

Attfield, Robin and Rebekah Humphreys (2017), 'Justice and non-human animals, Part II', *Bangladesh Journal of Bioethics*, 8(1), 44–57.

Barry, Brian (1978), 'Circumstances of justice and future generations', in R.I. Sikora and Brian Barry (eds), *Obligations to Future Generations*, Philadelphia, PA, USA: Temple University Press, 204–48.

Barry, Brian (1999), 'Sustainability and intergenerational justice', in Andrew Dobson (ed.), *Fairness and Futurity: Essays on Environmental Sustainability and Social Justice*, Oxford: Oxford University Press, 93–117.

First National People of Color Environmental Leadership Summit (1991), 'The principles of environmental justice', Washington, DC, USA, accessed 26 February 2018 at https://www.nrdc.org/resources/principles-environmental-justice-ej.

Harris, Paul G. (ed.) (2011), *China's Responsibility for Climate Change: Ethics, Fairness and Environmental Policy*, Bristol: Policy Press/Bristol University Press.

Jonas, Hans (1984), *The Imperative of Responsibility*, trans. Hans Jonas and David Herr, Chicago, IL, USA and London, UK: University of Chicago Press.

Kelbessa, Workineh (2012), 'Environmental injustice in Africa', *Contemporary Pragmatism*, 9(1), 99–132.

Möllendorf, Darrel (2002), *Cosmopolitan Justice*, Boulder, CO: Westview Press.

O'Neill, Daniel W., Andrew L. Fanning, William F. Lamb and Julia K. Steinberger (2018), 'A good life for all within planetary boundaries', *Nature Sustainability*, 1(2018), 88–95.

Parfit, Derek (1984), *Reasons and Persons*, Oxford: Clarendon Press.

Rawls, J. (1971), *A Theory of Justice*, revised edn (1999), Oxford: Oxford University Press.

Rawls, John (1993), *Political Liberalism*, New York, USA and Chichester, UK: Columbia University Press.

Scanlon, T.M. (1998), *What We Owe to Each Other*, Cambridge, MA, USA and London, UK: The Belknap Press of Harvard University Press.

Schopenhauer, Arthur (1851), *Parerga and Paralipomena: Short Philosophical Essays*, trans. Eric Francis and Jules Payne, Oxford: Clarendon Press (1974).

Singer, Peter (1979), *Practical Ethics*, 2nd edn (1993), Cambridge: Cambridge University Press.

Sterba, James (1998), *Justice for Here and Now*, New York: Cambridge University Press.

Sweetlove, Lee (2011), *Nature* (News), accessed 31 January 2019 at https://www.nature.com/news/2011/110823/full/news.2011.498.html.

United Nations (2015), *Sustainable Development Goals: 17 Goals to Transform Our World*, accessed 11 April 2017 at www.un.org/sustainabledevelopment/sustainable-development-goals/.

United Nations Framework Convention on Climate Change (2018), Statement on the Summary, accessed 31 January 2019 at https://unfccc.int/news/unfccc-secretariat-welcomes-ipcc-s-global-warming-of-15degc-report.

3 Common but differentiated responsibilities: agency in climate justice

Ivo Wallimann-Helmer

Ethical challenges concerning climate change often involve two issues that are closely connected. The first involves the just distribution of entitlements and burdens, and the second concerns the fair differentiation of responsibilities. The fairness of a distribution of entitlements and burdens can be assessed by relying on principles of climate justice. Although the fairness of any differentiation of responsibilities must rely on these principles, their applicability and the demands that they make depend strongly on three factors: (1) the policy domains at issue, including mitigation, adaptation and loss and damage; (2) the agents – most often states or other communities – bearing the responsibilities; and (3) the policy levels of international, regional, national or local climate action. Not all agents can be ascribed the same responsibilities, and not all measures for climate action can or should be realized by the same differentiation of responsibilities.

In this chapter, I examine how the differentiation of responsibilities for climate action depends on the domain and level of climate policy. The responsibility bearers may change depending on whether mitigation, adaptation or loss and damage policy are at issue. Since the responsibility bearers are embedded in complex nets of responsibilities, the domain and level of climate policy also define various institutions and principles of accountability. The principle of *common but differentiated responsibilities* is often the starting point of climate justice. But it also shapes what demands for climate action are most appropriate for the different agents responsible for climate action.

The chapter is structured according to the four aspects of responsibility (Bayertz 1995, Wallimann-Helmer 2016): (1) someone, the subject of responsibility, is always responsible for (2) something, the object of responsibility, (3) answerable to some institution, and (4) held accountable to some norm. I first explain why the differentiation of responsibilities between subjects varies depending on the domain of climate policy at issue. This variation is because, depending on the climate policy domain, the object of responsibility defines other agents as the most appropriate responsibility bearers. Next, I elaborate on states as collective agents and discuss why, depending on the nature of these subjects of responsibility, other demands for climate action are legitimate. The conditions necessary for a collective agent to be capable of taking on responsibility constrain what climate action can be expected. Since agents are always answerable to some (higher-level) institutions, and because in international politics these institutions themselves bear various responsibilities, climate action entails a complex net of responsibilities. I explore the consequences of this embeddedness for subjects of responsibility and the climate action that those subjects can be expected to take. My considerations in this chapter cannot decide the principles of climate justice relevant to international climate politics. However, they can help to explain why principles of climate justice cannot be assigned the same weight and relevance in all domains and at all levels of climate policy irrespective of the responsibility bearers being considered.

Subjects of responsibility and domains of climate policy

Under the umbrella of the United Nations Framework Convention on Climate Change (UNFCCC), there are three main domains of climate policy: mitigation, adaptation, and loss and damage (L&D). The most obvious subjects of responsibility in international climate politics are the state parties to the UNFCCC. They have agreed to a common responsibility of avoiding or minimizing dangerous climate change and its negative impacts. The first principle in Article 3 of the UNFCCC holds that all parties to the convention have common but differentiated responsibilities to protect the climate system (United Nations 1992):

> *The Parties should protect the climate system for the benefit of present and future genera-tions of humankind, on the basis of equity and in accordance with their common but dif-ferentiated responsibilities and respective capabilities. Accordingly, the developed country Parties should take the lead in combating climate change and the adverse effects thereof.*

According to this principle, the parties have accepted to share the burdens of mitigating climate change. In later decisions, the parties adopted the same principle to share the burdens for adaptation and L&D action (United Nations 2011, 2014). How these burdens must be shared among the parties to be fair is a typical question of justice (Gardiner et al. 2010).

However, justice is not only about distributing burdens. It also concerns the just distribution of entitlements. In the case of climate change, these entitlements not only include financing measures for adaptation and L&D but also rights to emitting greenhouse gases. The distribution of these entitlements also concerns justice (e.g. Moellendorf 2011). But, as I have argued elsewhere (Wallimann-Helmer 2019), the principles of justice defining a just distribution of entitlements need not be the same as those used to distribute the burdens of climate action. The same holds for the sub-jects of responsibility most plausibly under duty to take specific climate action in the three domains of climate policy. As I explain below, this is because different objects of responsibility define different agents as the most appropriate to take action. For example, demanding drastic reductions of emissions is not appropriate for states that can barely secure sufficiency levels for their citizens. On the other hand, given appropriate resources and infrastructure, these states might be best positioned to implement and maintain adaptation measures (Wallimann-Helmer 2016).

An example can help explain this point. Imagine that together with two friends, you are organizing a birthday party for a fourth friend. When it comes to the question of who bakes the birthday cake, it is very probable that you will let whoever of the three of you is the most experienced baker do the baking. However, the ingredients can be bought by whoever knows of a shop nearby with cake ingredients and owns a car. Perhaps one friend both bakes and shops, or two of your friends do one of these tasks each. Maybe, after these two tasks are allocated, you should volunteer to do other things. When discussing the distribution of tasks, all of you might bring forward distinct criteria or principles of justice. But you might also agree that it is most appropriate if the different tasks are done by those best able to do so.

This is a way of differentiating responsibilities that relies on criteria not of fairness but of efficiency. The subjects of responsibility for specific tasks should be those who are most capable of their completion. Arguing from this perspective means considering the various objects of responsibility and how well they fit the varying capacities of the three friends to organize the birthday party most efficiently. However, the subjects of responsibility might change for the distribution of the cake and other tasks at the party. Consequently, the friend most able to serve the cake will be the one responsible for doing so, the one with the most experience in mixing drinks will mix, and the one best able to lift heavy loads will carry empty bottles and bring new beer from the cooler.

In my view, this example most clearly illustrates the cases of adaptation and L&D policy. Both policy domains concern measures to minimize the risks of near-term climate impacts that cannot be prevented by mitigation efforts. The objects of responsibility in these domains are the measures to be taken to minimize risks of negative climate impacts (Wallimann-Helmer et al. 2019). Responsibilities must be differentiated for financing, implementing and maintaining these measures. It seems plausible to assign these responsibilities irrespective of contributions to climate change by following efficiency criteria. Those best able to finance, implement or maintain these measures should bear these responsibilities. Due to their affluence, developed countries are most plausibly demanded to provide financial assistance. Those states with greatest experience in taking adaptation action should help implement these measures where assistance is needed. Empirical research has shown that maintenance of adaptation is most sustainable if local communities are directly involved (Kaswan 2016, Nalau et al. 2015).

This approach to differentiating responsibilities for adaptation and L&D policy would treat the objects of adaptation and L&D as a kind of climate action that needs to be done most efficiently. However, international climate politics often treats the object of responsibility for adaptation and L&D differently (Huggel et al. 2016). Rather than focusing on the necessity of taking action, negotiations focus on the greater contributions of developed countries to changing climatic conditions.

In this debate, what becomes the object of responsibility is the historical and current emissions of the various parties. In consequence, those who have made greater contributions to the fact that adaptation and L&D action have actually become necessary are selected as responsible for contributing more to them. This significantly changes the whole picture of how responsibilities should be differentiated. Assigning responsibilities for efficient adaptation and L&D action becomes of secondary importance.

Recognition 'of contributing to climate threats is what this kind of approach prioritizes, but it misses what I take to be the main goal in this policy domain (Wallimann-Helmer 2015, 2016): immediate action for adaptation and L&D. To me, viewing emissions as the relevant object of responsibility is more plausible in the case of mitigation. To avoid dangerous climate change, the volume of emissions that can be produced is restricted to a predefined budget. The total sum of cumulative emissions that leaves the average global mean temperature below this target establishes an emissions budget to be distributed between all the parties. This budget is the cake that becomes relevant to the differentiation of responsibilities. The object of the common responsibility is to keep accumulated global emissions within this budget. The objects of the differentiated responsibilities are the emissions of the parties or their emission cuts.

It seems that the goal of the different policy domains of mitigation, adaptation or L&D define the criteria by which the objects of responsibility in climate action are most plausibly conceived. In the cases of adaptation and L&D, the goal most plausibly seems to be implementing effective measures to reduce negative climate impacts. In the case of mitigation, the goal is to keep accumulated global emissions within a predefined budget. However, how the appropriate criteria can be derived from the goal of a policy domain to describe the objects of responsibility correctly must remain a question to be researched. It seems to me that a more principled approach is needed.

The argument thus far has hinted at how the objects of responsibility define what subjects are most appropriate to responsibility. Those agents should be bearing responsibility that can best meet the criteria employed to conceive the objects of responsibility. In the example of the birthday party, I used efficiency criteria to explain this point. However, in other cases it might be other criteria, such as past experience, cultural ties or financial liquidity. In the next two sections, I explore the sense in which the capacities of states to be responsible collective agents, and their embeddedness in nets of responsibility bearers, define how they can be expected to take on responsibility for climate action.

Capacity and the nature of states as responsibility bearers

The most obvious subjects of the common but differentiated responsibilities under the UNFCCC are its parties: the sovereign states signing the convention. States as responsibility bearers are collectives that, like individuals, can lose their capacity to bear responsibility. Since to be effective today global climate action arguably must rely on sovereign states, it is crucial that they remain agents capable of responsibility (Wallimann-Helmer 2017). Otherwise they lose their capacity to take effective climate action. In order to understand which conditions can lead states to lose their capacity as responsibility bearers, the extent to which they can be grasped as collective agents has to be clarified. In what follows, I explore the conditions of agency for states. I show how the nature of states as collective agents is relevant for the normative expectations regarding their responsibilities for climate action.

Since states have organizational structures analogous to those of companies, considerations about corporate responsibility can be usefully applied to states. In business ethics, the extent to which corporations can be viewed as collective agents is a matter of intense debate. Slightly simplifying the debate about corporate responsibility, it is possible to distinguish two perspectives on states as subjects of responsibility. According to the first perspective, states can only be deemed subjects of responsibility if they can produce decisions within the framework of their governance institutions (Pasternak 2013). Collective agents are viewed as non-natural persons who, mediated by their decision-making structures and the individuals incorporated within them, can form intentions, generate knowledge, and reach and implement decisions. They can therefore assume responsibility, like natural human beings, and are subject to the same conditions of agency (French 1984).

In contrast, the second perspective assumes that states are not agents like natural persons. Collective agents are only persons in a derivative sense because the status of agency is assigned either externally or through their members (Werhane 1985). This status is assigned to states because their political decisions and actions have an impact on citizens and subnational organizational structures, and on governance structures and individuals external to them (Miller 2007). The international community assigns states responsibility for their actions within the framework of international treaties. Citizens view their own state as having responsibility for realizing certain goals in their names.

Two essential differences between companies and states should be taken into account before deciding that the responsibility of states as collective agents resembles those of companies (Wall 2001). First, employees are free to leave a company at any time and usually join companies voluntarily. In contrast, citizens are usually members of a state involuntarily, and leaving or joining a state can only be achieved at considerable expense (if at all). Second, states can only embrace their responsibilities and action if they pass on any concomitant costs to their citizens. For example, most states generate the capital they need to balance climate action costs not through the buying and selling of goods on the open market, like companies, but by levying taxes. As with any other law to achieve climate goals, this is an impairment of individual liberties and places burdens on citizens. For these two reasons, it is therefore crucial to clarify the extent to which citizens have to support the responsibilities assumed by their states for those states to remain collective agents.

Depending on which of the two perspectives of collective responsibility forms the basis for a state being deemed a collective agent, different conditions apply to how citizens have to back the action of their states. If states act like natural persons through their citizens and institutions, citizens should be able to influence state activities to a sufficient extent (Stilz 2011). If the status of collective agency is only assigned to states as secondary persons, it is more relevant that the global community acknowledges a state as a responsible agent and that its citizens can identify with it as a subject of collective responsibility acting on their behalf (Jubb 2014). Correspondingly, citizen participation in a democratic sense is only a key prerequisite for the capacity of a state to be a responsible agent from the first perspective. In the second perspective, a state also has a capacity for responsibility if the international community assigns this agency and if its citizens can sufficiently identify with its policy action. The ability of citizens to influence state activity through institutional decision structures is less important or not important at all.

However, it is implausible to assume that a state's capacity for responsibility is constituted solely through the sufficient involvement of all its citizens, or solely through their identification with the actions of their state. An appropriate conception of states as collectives with the capacity for responsibility has to incorporate both perspectives of collective responsibility, at least to a certain degree (cf. Parrish 2009). The degree of citizen involvement in the collective decision-making of a state is a key feature by which to assess what can be expected from states with regard to climate action. Since states become established over the course of long historical development, their capacity for responsibility strongly depends on their historically developed self-image as collective agents acting on behalf of their citizens (Hampton 1997, Wallimann-Helmer 2013). Citizens of a state will only be prepared to back and support the policy action of their state to the extent that action is compatible with how they conceive of their state as a subject of responsibility.

Consequently, if more is demanded of states than their citizens find compatible with their conception of their state as an agent taking on responsibility, there is a high risk that these states will lack sufficient support for their climate action and so lose their capacity to act as responsible agents. There is a real danger that states will lose that capacity if their citizens do not properly support their actions. Political stalemates and blockades are the least serious consequences. Nationwide protests, riots and even civil wars would be more devastating.

Whether more or less drastic, none of these would be in the interests of timely and effective climate action. As a consequence, states should only be deemed responsible for climate action inasmuch as what is demanded can be aligned with their historically developed self-image as collective responsible agents. If this socio-ontological thesis is correct, then the nature of states as collective agents defines constraints on what can be expected from them in the three domains of climate policy. Their responsibilities should not be greater than their citizens expect them to be. Expecting more would mean that the most crucial agents in international climate policy may lose their capacity for agency, and hence that the necessary climate action might not be taken.

Therefore, whatever differentiation of responsibilities is deemed most appropriate to achieve objects of responsibility, to ensure that the main actors in international climate politics are and remain agents capable of taking on responsibility, it is necessary to consider the nature of each state before assigning responsibilities. Since in international climate politics it is not only states that are responsibility bearers, but also supranational and subnational bodies of governance, the same holds for these kinds of collective agents. However, whilst my distinction between the two perspectives on states as collective agents already has been simplified, there is a lack of conceptual research into these kinds of supranational and subnational collective agents and the conditions required for them to remain functioning subjects of responsibilities. Moreover, further empirical investigation is needed into whether my socio-ontological thesis bears some relevance for the possibility of collective agency at all. The next section sheds some light on the conceptualization of how to understand supranational and subnational entities as collective agents.

Institutions of answerability and complex nets of responsibilities

I have argued that it is crucial to consider the nature of states and other collective agents relevant for climate action before assigning responsibilities. The nature of these agents shows the conditions under which they can bear responsibility for whatever policy object they are assigned. However, the institutions of answerability, the third aspect of the concept of responsibility, can also constrain whether an agent lives up to its responsibilities on several grounds. In this section, I explore several implications for national, supranational and subnational bodies of governance as subjects of responsibilities answerable to various higher-level institutions, such as the international Green Climate Fund, the Alliance of Small Island States, the UNFCCC or, in the case of subnational actors, states themselves. I argue that the relevant collective agents in international climate politics are embedded in complex nets of responsibilities that shape what climate action can be expected from them.

To better understand why I believe this to be the case, it is helpful to look more closely at the structure of international climate politics. International climate politics is a system with at least two levels of policy-making (Miller 2008, Wallimann-Helmer 2019). At the first, international level, the parties to the UNFCCC have to negotiate and accept differentiations of responsibilities among them. Due to the lack of an international institution to enforce these agreements, lack of common acknowledgement of an agreement leads either to non-compliance or to opting-out by some states. This is an important reason why at the second, domestic level, national self-determination in the implementation of internationally agreed responsibilities should be guaranteed. Otherwise, there is a high risk that states lose their capacities to be responsible agents. Demanding that a state implements policies not legitimized by its decision structures means challenging its historically developed self-conception of how it should bear responsibility on behalf of its citizens. Depending on how severely this self-conception is undermined, this can mean that states cease to be agents capable of responsibility and hence do not realize the climate action assigned to them.

However, there are not just two levels of governance in international climate politics; many additional agents, at various policy levels, can be deemed responsible, especially in the case of adaptation policy (Blackburn and Pelling 2018, Roggero et al. 2018, Nalau et al. 2015). As already mentioned, both states and subnational and supranational bodies of governance can be assigned varying responsibilities. None of these potentially relevant subjects of responsibility exist in isolation from each other; their individual responsibilities are part of a complex net of responsibility dependencies. These dependencies exist because, in many cases, higher-level subjects of responsibility are the institutions of answerability for lower-level subjects of responsibility. At the level of international climate politics, countries are answerable to the institutions established under the umbrella of the UNFCCC. Some states are also answerable to regional structures or other coalitions, such as those that are members of the Alliance of the Small Island States. At the domestic level, governments and subnational entities are answerable to their people, but also to their states' various institutions. And individual citizens are accountable to national law and their social communities.

How these relations of answerability are shaped depends on the nature of the institutions considered. It depends both on the nature of the subjects of responsibility and on the support and demands these subjects experience from their various institutions of answerability. For example, institutions like the Green Climate Fund have been established to finance corresponding climate action. Those receiving these funds are answerable to the funding institutions and must prove that the money has been invested appropriately. However, if the bearers of adaptation responsibilities do not receive appropriate support for adaptation action, they may not be equipped to meet their responsibilities effectively. This is also true if institutions of accountability do not allow subjects to meet their responsibilities in line with the goal of the climate policy domain at issue or if different institutions of accountability demand the fulfilment of conflicting goals. Whilst this second kind of constraint is more drastic for the capacity of an agent to take climate action, the first is more frequent in climate policy. Often there is a lack of support, as in the domain of adaptation finance. Many states are not properly assisted in competence building and do not have sufficient resources or the necessary technology to implement the measures required (Huggel et al. 2016).

Since the responsibility bearers are embedded in complex nets of responsibilities, how different institutions shape conditions of answerability and support the fulfilment of responsibilities by lower-level agents also defines the capacities of responsibility bearers. If higher-level institutions demand more from lower-level subjects than accords with their self-conception as responsible agents, it seems plausible that less burdensome climate action can be expected from them. Excessive demands seem to weaken capacities for responsibility. The same seems to be true when agents lack appropriate support. In such cases, many would argue that responsibilities vanish when necessary support is missing. And if fulfilment of conflicting responsibilities is requested by institutions of accountability, it seems clear that it would be unfair to demand the fulfilment of all of them anyway. It seems to be unfair to expect a state with low technological development not only to increase its economic rate of growth for better climate resilience but at the same time asking it to reduce its emissions without first providing transfer of clean technology.

Here these considerations must remain as tentative as they are. It is necessary to further explore how higher-level subjects of responsibility can be viewed as institutions of answerability. This means analysing the conditions under which the subjects of responsibility at different levels of climate policy can be conceived as collective agents of responsibility at all. In addition, further investigation is needed into how constraining conditions, due to institutions of answerability, change the responsibilities of responsibility bearers. The conditions under which the responsibilities of these subjects are legitimately conceived as weakened need to be clarified.

Conclusion

I have identified three gaps in normative research on climate justice. First, the domain of climate policy defines what kinds of agents should bear responsibility for achieving the key policy goal in this area. The goal of a policy domain as the object of differentiated responsibilities defines the agents most appropriate to take climate action. However, a more principled approach is needed to clarify how a goal in a policy area is best described and what this means for defining the appropriate agents that bear responsibility. Second, the very nature of agents of responsibility defines what can legitimately be expected of them. How agents become responsibility bearers defines their capacities as responsible agents. With regard to states as the main actors in international climate politics, their historically developed self-conception as collective agents is crucial to understanding their capacities to bear various responsibilities. Indeed, it is necessary to research empirically whether this socio-ontological hypothesis has relevance and how collective agents other than states must be conceptualized as bearers of differentiated responsibilities. Third, agents in international climate politics are embedded in complex nets of responsibilities. This is because the institutions of answerability for some lower-level responsibility bearers are simultaneously subjects of responsibility answerable to higher-level institutions. What this embeddedness means for the responsibilities of single collective agents at various levels of climate policy is the final research gap that this chapter identifies.

Beyond these three research gaps, when trying to realize climate justice, it is crucial to consider the agents and their dependences when assigning differentiated responsibilities. However, this leaves entirely open what differentiation of responsibilities would be fair. It only demands that the various capacities of agents are considered when differentiating responsibilities for climate action. Although considering the capacities of agents when distributing responsibilities invokes the ability-to-pay principle of climate justice, it is not clear whether taking into account only this principle of justice meets our intuitions about justice. Many would argue that it is crucial that all agents take on burdens of climate action in proportion to their contribution to climate change or in proportion to their benefits from others' contributions. Others believe that it is crucial to fulfil our duties towards future generations, whatever restrictions this might mean for us living today.

How the relation between these potentially conflicting demands of justice plays out in the various domains and levels of climate policy is not clear. It is an open question whether ensuring the capacities of collective agents to function as responsibility bearers always has priority or whether justice considerations are more important. Clarifying how responsibility is to be understood and differentiated for the various domains and levels of climate policy cannot help in deciding this question fully, but it reminds us to be cautious when assigning responsibilities. For immediate climate action to be effective, and thereby to realize climate justice, it is crucial that the bearers of responsibility remain functioning collective agents.

Acknowledgement

As always, I would like to thank Simon Milligan for his very much appreciated feedback on earlier versions of this chapter.

References

Bayertz, Kurt (1995), 'Eine kurze Geschichte der Herkunft der Verantwortung' [A brief history of the origin of responsibility], in K. Bayertz (ed.), *Verantwortung: Prinzip oder Problem? [Responsibility: Principle or Problem?]*, Darmstadt: Wissenschaftliche Buchgesellschaft, pp. 3–71.
Blackburn, Sophie and Mark Pelling (2018), 'The political impacts of adaptation actions: social contracts, a research agenda', *Wiley Interdisciplinary Reviews: Climate Change*, 9(6), e549.
French, Peter (1984), *Collective and Corporate Responsibility*, New York: Columbia University Press.
Gardiner, Stephen M., Simon Caney, Dale Jamieson and Henry Shue (eds) (2010), *Climate Ethics: Essential Readings*, Oxford, UK and New York, NY, USA: Oxford University Press.
Hampton, Jean (1997), *Political Philosophy, Dimensions of Philosophy Series*, Boulder, CO: Westview Press.
Huggel, Christian, Ivo Wallimann-Helmer, Dáithí Stone and Wolfgang Cramer (2016), 'Reconciling justice and attribution research to advance climate policy', *Nature Climate Change*, 6(10), 901–908.
Jubb, Robert (2014), 'Participation in and responsibility for state injustices', *Social Theory and Practice*, 40(1), 51–72.
Kaswan, Alice (2016), 'Climate change adaptation and theories of justice', *Archiv für Rechts- und Sozialphilosophie [Archive for Philosophy of Law and Social Philosophy]*, Beihefte No. 149, pp. 97–118.
Miller, David (2007), *National Responsibility and Global Justice*, Oxford: Oxford University Press.

Miller, David (2008), 'Global justice and climate change: how should responsibilities be distributed?', *The Tanner Lectures on Human Values*, **28**, 117–56.

Moellendorf, Darrel (2011), 'Common atmospheric ownership and equal emissions entitlements', in D.G. Arnold (ed.), *The Ethics of Global Climate Change*, Cambridge: Cambridge University Press, pp. 104–23.

Nalau, Johanna, Benjamin L. Preston and Megan C. Maloney (2015), 'Is adaptation a local responsibility?', *Environmental Science & Policy*, **48**, 89–98.

Parrish, John M. (2009), 'Collective responsibility and the state', *International Theory*, **1**(1), 119–54.

Pasternak, Avia (2013), 'Limiting states' corporate responsibility', *Journal of Political Philosophy*, **21**(4), 361–81.

Roggero, Matteo, Alexander Bisaro and Sergio Villamayor-Tomas (2018), 'Institutions in the climate adaptation literature: a systematic literature review through the lens of the Institutional Analysis and Development framework', *Journal of Institutional Economics*, **14**(03), 423–48.

Stilz, Anna (2011), 'Collective responsibility and the state', *Journal of Political Philosophy*, **19**(2), 190–208.

United Nations (1992), *Framework Convention on Climate Change*, accessed 9 March 2018 at: http://unfccc.int/files/essential_background/convention/background/application/pdf/convention_text_with_annexes_english_for_posting.pdf.

United Nations (2011), *Report of the Conference of the Parties on its Sixteenth Session, held in Cancun from 29 November to 10 December 2010: Part Two: Action taken by the Conference of the Parties at its Sixteenth Session* (accessed 8 June 2015).

United Nations (2014), *Report of the Conference of the Parties on its Nineteenth Session, held in Warsaw from 11 to 23 November 2013: Part Two: Action taken by the Conference of the Parties at its Nineteenth Session* (accessed 3 September 2014).

Wall, Steven (2001), 'Neutrality and responsibility', *Journal of Philosophy*, **98**(8), 389–410.

Wallimann-Helmer, Ivo (2013), 'The Republican tragedy of the commons: the inefficiency of democracy in the light of climate change', *Ancilla Iuris*, pp. 1–14.

Wallimann-Helmer, Ivo (2015), 'Justice for climate loss and damage', *Climatic Change*, **133**(3), 469–80.

Wallimann-Helmer, Ivo (2016), 'Differentiating responsibilities for climate change adaptation', *Archiv für Rechts- und Sozialphilosophie [Archive for Philosophy of Law and Social Philosophy]*, Beihefte No. 149, pp. 119–32.

Wallimann-Helmer, Ivo (2017), 'Kollektive Verantwortung für den Klimaschutz' [Collective responsibility for climate protection], *Zeitschrift für Praktische Philosophie [Journal for Practical Philosophy]*, **4**(1), 211–38.

Wallimann-Helmer, Ivo (2019), 'Justice in managing global climate change', in T. Letcher (ed.), *Managing Global Warming: An Interface of Technology and Human Issues*, London, UK, San Diego, CA, USA and Cambridge, MA, USA: Academic Press, pp. 751–68.

Wallimann-Helmer, I., Meyer, L., Mintz-Woo, K., Schinko, T. and Serdeczny, O. (2019). 'The ethical challenges in the context of climate loss and damage', in R. Mechler, L. M. Bouwer, T. Schinko, S. Surminski, & J. Linnerooth-Bayer (eds.), *Climate Risk Management, Policy and Governance. Loss and damage from climate change: concepts, methods and policy options*, Cham: Springer, pp. 39–62.

Werhane, Patricia Hogue (1985), *Persons, Rights, and Corporations*, Englewood Cliffs, NJ: Prentice Hall.

4   The world as it is: a vision for a social science (and policy) turn in climate justice

David E. Storey

In 2006, philosopher Robert Frodeman (2006) published an article titled 'The policy turn in environmental philosophy'. He argued that environmental ethicists had become too narrowly focused on abstract meta-ethical issues about, for example, the intrinsic value of nature, and too detached from the real-world economic, cultural and political contexts in which environmental problems occur. I argue that we require a similar turn in how we think about and work for climate justice. The difference is that while environmental ethics has been dominated by an ecological moral idealism, climate ethics has been dominated by a humanitarian moral idealism. The *lingua franca* of climate ethics is overwhelmingly cosmopolitan (see Harris 2011, 2016). Concepts like rights, duties, equality, justice and so on are often assumed to apply to all people at all places and times, regardless of 'tribe'. My argument is not that cosmopolitan forms of climate justice are wrong – indeed, they are probably the most defensible – but that, given the failure of ethics to penetrate climate policy, and of climate policy to gain political traction, perhaps we ought to stop asking whether our accounts of climate justice are 'right or wrong' and start asking why they have not worked so far.

In this chapter, I lay out a research agenda that brings together moral philosophy and the social sciences to develop a more effective approach to climate justice. First, I briefly survey the state of climate change ethics in academic philosophy. My aim here is not to quibble with the merits of any particular argument. Instead, I want to point out that the focus of climate ethics has been on minimizing and preventing harm, human rights violations, and distributive and corrective justice, often with respect to future generations and poor people in developing countries. One reason such arguments have not penetrated climate policy and politics is that they do not resonate with the major moral concerns of many people, particularly Americans, and especially conservative Americans. By 'conservative', I mean two distinct but often overlapping ideological camps: 'economic' conservatives who favour low taxes and minimal government regulation, and 'social' conservatives who hold traditional views on issues such as abortion and same-sex marriage and are highly religious. While my argument could be applied globally, I focus on the United States because its leadership is probably essential to successful and impactful global coordination on climate policy, and because its political culture is uniquely resistant to federal action on climate change.

Second, I present frameworks from anthropology (Mary Douglas's grid-group theory), social psychology (Jonathan Haidt's moral foundations theory) and political science (Ronald Inglehart's World Values Survey) and explain how they can help us make sense of the failure of climate justice so far. Given the popular resistance and indifference to action on climate change – and, by extension, the failure of arguments by philosophers, policy makers, politicians and scientists – we must think through the normative implications of research in the social sciences that is telling us so much about people's beliefs, values, judgements and perceptions around climate change. Scholars and activists can and should draw on frameworks from different disciplines to develop a flexible guide for multiple moral framings of climate justice.

Throughout, I identify three key worldviews – traditional, modern and postmodern – that fuel gridlock on climate change. Put simply, the climate justice debate has been framed as a zero-sum conflict between moderns and postmoderns that ignores traditionalists. Put another way, such framing needlessly treats two worldviews as incompatible and pretends that the third does not exist. We need to seek ways of framing climate justice and communicating climate policy that aim at convergence and congruence between the major worldviews. If we are to succeed in bending the moral arc of history toward climate justice – to remake the world as it ought to be – we need to do a better job of working with the world as it is.

The narrow frame of climate justice

I think it is fair to say that moral and political philosophy of climate change is in its early days, but one thing seems clear: the focus is on developing a theory of global, environmental and intergenerational justice. The harms likely to be caused by climate change are many: from sea-level rise, flooding and extreme weather causing damage to agriculture and infrastructure; from drought impacting crop yields and food security; from heat waves; from water- and vector-borne diseases; loss of life due to all of the above; and so on. These harms will fall disproportionately on future generations and poor people in developing countries. Many thinkers have attempted to identify the rights and duties of individuals, businesses and states that are relevant to the climate problem. Questions of distributive, procedural and corrective justice abound, especially who ought to bear the costs of mitigation, adaptation and, most controversially, compensation for loss and damage. British economist Nicholas Stern (2010) has based his case for climate action on cost–benefit analysis, an essentially utilitarian form of justification; similarly, moral philosopher John Broome (2012) has made the case for climate action on utilitarian grounds. Simon Caney (2010) has argued that we ought to view climate change through the lens of human rights, claiming that climate change poses threats to rights to life, health and subsistence (and perhaps rights to economic development and to not be displaced). Darrel Moellendorf (2015) has argued for a right to sustainable development. Henry Shue (2010), also arguing from a rights perspective, has relied on the 'polluter pays' principle, claiming that we should follow a principle of historical accountability for past emissions. Appealing to welfare, as well as to political and economic feasibility constraints, Peter Singer (2010) disagrees with any such historical principle, and he has proposed equal per capita emissions and an equal share of entitlements to the global atmospheric sink.

This is not an exhaustive list but a representative one, and my aim is not to discuss the strengths and weaknesses of these various approaches. I merely want to highlight a feature of what many of them share. First, though their rationales differ, their conclusions are roughly the same. As Matthias Fritsch notes,

> some philosophers have argued that the case for strong and immediate action on climate change is ethically overdetermined and that there are multiple compelling arguments for the same conclusion – that the US and other rich industrialized countries have a moral obligation to take the lead in what ultimately needs to be a global effort to reduce emissions. (Fritsch 2012: 225)

Put simply, most ethical analyses of climate change conclude that the rich countries ought to be doing far more than they currently are committed to do under the Paris Agreement to reduce their emissions and invest in adaptation efforts in developing countries. Donald Brown concurs that there is overlapping consensus within climate ethics, suggesting that, contra Stephen Gardiner's (2011) claim that we our plagued by a 'theoretical storm' in which our extant moral theories leave us stranded, such theories can, in fact, offer useful guidance:

> The first-order problem for climate change ethics is not to determine finally what ethics requires on all issues, but to help people see that positions of contending parties are unjust. That is, a first order problem is to get ethics considered at all in the climate change debate, not to get agreement on what ethics requires. (Brown 2013: 247–8)

It is this first-order problem with which I am here concerned.

This problem raises a simple question: why has ethics failed to make a major dent in the climate change debate? To be fair, ethical considerations have not been completely ignored: Shue and Broome, for instance, have contributed to climate documents produced by the United Nations and the Intergovernmental Panel on Climate Change. However, I also think it is fair to say that the impact of ethics on the debate has fallen short of what most ethicists would deem adequate. Brown offers several reasons, one of which I want to point out. Ethics has been sidelined in climate policy discussions because it has been deemed abstract, idealistic and detached from political and economic feasibility constraints. The title of an article by Eric Posner (2013), a proponent of what Gardiner calls the 'pure policy approach', sums it up: 'You can have either climate justice or a climate treaty, but not both'. The problem with this way of thinking, as Gardiner points out, is that 'It frames climate change as a "pure policy" problem, where the relevant question is "what works" and the answers must come from science, economics, international relations and related disciplines...climate change becomes a matter best left to the technocrats' (Gardiner and Weisbach 2016: 46–7). The effect of such framing, however, is to conceal the value judgements essential to climate policy, and out-source ethical decision making to experts who hail from purportedly value-neutral disciplines. I thus agree with Brown's call for 'an applied environmental ethics that examines specific scientific and economic arguments and their erroneously assumed "value-neutral" underpinnings if ethicists hope to influence policy forma-tion' (Brown 2013: 15–16).

I think that part of the reason ethics has failed to influence climate policy is that most ethical analysis tends to focus on rights, justice and welfare. This would not seem to be a problem – what else would they focus on? However, what I want to suggest is that this actually is a problem. Climate ethics has framed the issue in moral categories that fail to motivate large groups of people. This blind spot – one that I think research in the social sciences can help to illuminate – is one reason that ethics has failed to influence climate policy.

Re-framing climate justice: integrating research from the social sciences

The United States can only make progress on climate change if the warring factions in the climate debate are *led by their own values and views to recognize, resonate with, and respond to the problem – to think, feel and act in new ways by drawing upon old values and views*. Both the problem and the solutions need to be framed in ways that are congruent with the major worldviews and value systems in the political culture of the United States. The climate problem is not going away, but neither are these worldviews. Facts and physics are stubborn things, yes, but values and views are stubborn, too. Rather than wringing our hands while waiting for nationalists to think globally and act locally, or for business folks to put planet over profit, we scholars of climate justice and activists need to get better at thinking and feeling into these other perspectives in order to speak their language. The following frameworks from the social sciences can help us to do so.

Grid-group theory

There is a widespread misconception that the obstacle to action on climate change is ignorance – that those who deny or do not care about climate change just 'don't know the facts', that we just need more information, data, science and so on. Yet there is fascinating research from the social sciences suggesting that the prob-lem is not scientific literacy, but differences in worldviews and values. In *Nature Climate Change*, Dan Kahan and his colleagues provide evidence that (1) cultural worldviews powerfully shape individuals' attitudes toward climate change and (2) increased scientific literacy and numeracy is not correlated with increased concern about climate change. Common sense supports the 'science comprehension thesis', which they describe thus: 'Because members of the public do not know what scientists know, or think the way scientists think, they predictably fail to take climate change as seriously as scientists believe they should' (Kahan et al. 2012: 732). On the other hand, the 'cultural cognition thesis' claims that 'individuals, as a result of a complex of psychological mechanisms, tend to form perceptions of societal risks that cohere with values characteristic of groups with which they identify' (Kahan et al. 2012: 732).

The researchers employ a model from the study of the cultural cognition of risk, which is based on the cultural theory of risk, or grid-group theory, first advanced by anthropologist Mary Douglas (1970). Douglas subsequently developed the theory with political scientist Aaron Wildavsky to determine how cultural worldviews affect individuals' perceptions of environmental and technological risks (1982). The theory plots individuals along two axes: 'grid' (vertical) and 'group' (horizontal). *Grid* measures the extent of individual variability and social stratification in a culture and runs along a spectrum of hierarchy (high) to egalitarianism (low). *Group* measures the strength of bonds between individuals in a culture and runs along a spectrum of individualism (low) to communitarianism (high). This produces four types of worldviews: fatalism, hierarchism, individualism and egalitarianism. Mike Hulme describes them thus:

> both hierarchists and egalitarians share a sense of solidarity as members of society; they are strongly group-oriented and feel bonded to larger social units. They differ, however, in whether they see this social bonding as primarily vertical or horizontal. The hierarchists see a strong social structure of rank, role and place, with social interaction governed by multiple sets of rules. The egalitarians view everyone as fundamentally equal, joined together through voluntary associations with few governing rules. The other two categories – individualists and fatalists. . .see little or very weak social bonding within society. The individualists additionally see little need for any social structuring of relationships around conventions or rules, whereas the fatalists accept their position as isolated individuals, but within a stratified and rule-bound society. (Hulme 2009: 186–7)

Buzz Holling (1986) and Michael Thompson (1998) subsequently developed four myths of nature that, as Hulme points out, map onto these four categories. Fatalists see nature as capricious and chaotic; hierarchists see it as perverse but tolerant, to some extent uncontrollable but stable within limits and safe given prudent stewardship; individualists see nature as benign, as something to be used for our benefit; and egalitarians see it as ephemeral and fragile, a delicate balance that humans must work to maintain (Hulme 2009: 188–90).

To test the 'cultural cognition thesis', Kahan et al. (2012) asked subjects who identified with different worldviews the question, 'How much risk do you believe climate change poses to human health, safety, or prosperity?', controlling for levels of scientific literacy and numeracy. They hypothesized that subjects with a 'hierarchical, individualistic' worldview that 'ties authority to conspicuous social rankings and eschews collective interference with the decisions of individuals possessing such authority' would be less concerned with climate change risks than those with an 'egalitarian, communitarian' worldview, 'one favouring less regimented forms of social organization and greater collective attention to individual needs' (Kahan et al. 2012: 733). And this is indeed what they found.

What they also found, however, is that increased scientific literacy and numeracy did *not* lead to an increase in concern with climate change among subjects with a hierarchical, individualistic worldview. In fact, the reverse happened: in response to the question, 'How much risk do you believe climate change poses to human health, safety, or prosperity?', egalitarian-communitarians' scores increased, but those of hierarchical-individualists *decreased*, again controlling for levels of scientific literacy and numeracy. In short, there is reason to believe that cultural types are sticky and resistant to attempts to change people's attitudes around climate change by appealing to facts, empirical evidence and scientific research. To get a better sense for the moral psychological makeup of individuals in these four types, we can turn to recent work in social psychology.

Moral foundations theory

Moral foundations theory (hereafter MFT), popularized by social psychologist Jonathan Haidt (2009), holds that human nature has evolved at least six basic moral foundations or 'taste buds': authority/subversion, loyalty/betrayal, sanctity/degradation, liberty/oppression, fairness/cheating and care/harm (brief descriptions of the foundations can be found at moralfoundations.org). The MFT has generated much research and debate, but my concern with the theory is its political application and normative implications. One of the most striking findings from research testing of the model is that there are clear relationships between which foundations a person values and their political orientation. While conservatives value all of the foundations, liberals – broadly construed as individuals who favour high government intervention in the economy to redress the social and environmental impacts of capitalism and are tolerant around social issues like abortion and same-sex marriage – score highly on only two foundations: care/harm and fairness/cheating (Haidt 2009). An important distinction here is that some foundations are 'individualizing', while some are 'binding'. As Annukka Vainio and Jaana-Piia Mäkiniemi explain, 'the care/harm and fairness/cheating foundations are referred to as individualizing foundations, with their emphasis on the rights and welfare of individuals. . . The three other foundations – loyalty/betrayal, authority/subversion and sanctity/degradation – are referred to as binding foundations and have an emphasis on group-binding loyalty, duty and self-control' (Vainio and Mäkiniemi 2016: 268).

What happens when we look at climate change through the lens of MFT? There has been some research done on this in the United States and Finland, and the findings can give us some indication for how we might use this knowledge to think strategically about framing climate justice. Researchers found evidence of 'the potential importance of moral foundations as drivers of intentions with respect to climate change action', which 'suggests that compassion, fairness and to a lesser extent, purity, are potential moral pathways for personal action on climate change in the USA' (Dickinson et al. 2016: 1). This last finding with respect to purity is important, because it is a potential pathway to engaging conservatives on climate change and because there is other research suggesting that the use of language that activates the sanctity/degradation foundation is associated with increased pro-environmental attitudes among conservatives.

These findings are also generally consistent with studies indicating that 'those who endorse general binding foundations do not perceive climate change as a moral issue' (Vainio and Mäkiniemi 2016: 273). And as the authors of another study note, 'where climate change is concerned, ethical discussions have focused mainly on nonharming and fairness, assuming that the other moral axes (those most prized by conservatives) are not relevant' (Dickinson et al. 2016: 3). As I suggested above, the lion's share of the literature in climate ethics adopts this approach, potentially alienating potential allies. How, then, might we activate the binding foundations to encourage conservatives to see the moral relevance of climate change, and be motivated to take action?

Hanno Sauer describes the logic of such an approach by distinguishing between 'strategic' and 'genuine' appreciation of moral foundations:

> *the strategic appreciation of conservative moral foundations occurs when liberals deliberately frame their concerns in terms which will resonate with the conservative, but which they would not prefer otherwise. One instructive example for this is climate change, where it can be shown that even though political conservatives are less likely to endorse measures against climate change, their position can be shifted in the direction of such measures if their importance is described in conservative terms. . .* (Sauer 2015: 157)

It turns out that the study described in 'How are moral foundations associated with climate-friendly consumption?' did just that. The study had subjects respond to questions about climate change that used language invoking each foundation. For instance, with respect to the authority/subversion foundation, they asked subjects to respond to the following statements:

- 'Climate change increases wars and conflicts among nations';
- 'Whether climate change mitigation is the duty of every citizen or not';
- 'The climate experts' recommendations ought to be followed'.

The following statements invoked the loyalty/betrayal foundation:

- 'Climate change mitigation is fighting for the fatherland';
- 'Whether climate change threatens the lives of [one's compatriots] or not';
- 'Whether climate change is a threat to my family or not'.

In contrast, the following statements invoked the sanctity/degradation foundation:

- 'Whether CC [climate change] is unnatural or not';
- 'Whether CC will increase the number of infectious diseases or not';
- 'CC will change [the] natural order created by God'.

The results of the study showed that both groups – those with individualizing foundations and those with binding foundations – saw climate change as a moral issue (Vainio and Mäkiniemi 2016: 276). While research using MFT is in its early stages, such results are promising. They indicate that framing climate change discussions using different moral categories can lead people who might otherwise not care about the issue to shift their attitude and take action.

Let me add one caveat about the use of MFT in this context. Haidt has argued that since conservatives have access to a wider set of moral foundations than liberals, since political disagreements are due to people's different moral foundations, and since such disagreements cannot be resolved through rational discourse, the burden is on liberals to try and find common ground. I think this is the wrong conclusion to draw from the research on MFT because it does not leave room for any notion of moral development tied to the effects of education and rational reflection. I suspect this is due to a social scientific preoccupation with description that leads Haidt to regard different moral psychologies as merely, well, different. However, as Sauer perceptively argues, if we examine research done on MFT, 'we see that a better education and sound debunking arguments leave "liberal" judgments about harmful acts or rights-violations unaffected, undermining only "conservative" moral beliefs, and that moral arguments based on liberal considerations often override those based on conservative ones' (Sauer 2015: 162). That is, there is evidence that conservatives can be led to revise their judgements based on rational testing, *but not the reverse*. This idea – that there is a certain developmental trajectory to types of values and worldviews – finds expression in the next framework that I examine.

The World Values Survey

The World Values Survey (WVS) is an ongoing project led by political scientist Ronald Inglehart. From 1981 to 2014, it collected nationally representative data from over 100 countries comprising over 90 per cent of the global population. The survey aims to identify how societies' values are shaped by political and economic conditions and, in particular, how and why major changes in values happen. From this massive data set, Inglehart has developed what he calls 'Evolutionary Modernization Theory', according to which 'economic and physical security are conducive to xenophobia, strong in-group solidarity, authoritarian politics and rigid adherence to their group's traditional cultural norms – and conversely that secure conditions lead to greater tolerance of outgroups, openness to new ideas and more egalitarian social norms' (Inglehart 2018: 8). Like grid-group theory, the WVS 'cultural map' plots countries along two axes: traditional values versus secular-rational values (vertical) and survival values versus self-expression values (horizontal). (The WVS map and analysis can be found at worldvaluessurvey.org.) I quote here the descriptions of the four types of values:

- *Survival values* place emphasis on economic and physical security. This is linked with a relatively ethnocentric outlook and low levels of trust and tolerance.
- *Traditional values* emphasize the importance of religion, parent–child ties, deference to authority and traditional family values. These societies have high levels of national pride and a nationalistic outlook.
- *Secular-rational values* have the opposite preferences to the traditional values. These societies place less emphasis on religion, traditional family values and authority.
- *Self-expression values* give high priority to environmental protection, growing tolerance of foreigners, gays and lesbians and gender equality, and rising demands for participation in decision-making in economic and political life. (World Values Survey 2014)

Having laid out the crux of the theory, let me make a few points to emphasize its relevance to climate justice. First, let us draw some connections to the prior frameworks. The four value types clearly correspond to the four grid-group identities: fatalists espouse survival values, collectivists embrace traditional values, individualists embrace secular-rational values, and egalitarians embrace post-materialist values. If this is so, we would expect them to have similarly corresponding views of nature: only egalitarians, who perceive nature as fragile, would see climate change as morally concerning. As for MFT, those with traditional values would clearly score high on the binding foundations of authority and loyalty, while those with secular-rational and/ or self-expression values would score higher on individuating foundations.

Second, I want to elaborate on three prominent value systems in the culture of the United States, which I will refer to as traditional, modern (secular-rational) and postmodern (self-expressive) in order to clarify the tensions between them and how these tensions fuel gridlock on climate change. Put simply, these three value systems have different approaches to climate justice. The traditional value system is ethnocentric. It sees the social world as a hierarchy of clearly prescribed roles governed by rigid rules. The cosmos is regarded as the ordered plan of a personal god who determines absolute values of good and evil. Traditionalists tend to see the world in black and white, us vs. them, absolutist terms. Nature is an order created by a benevolent god, and is there for human use, and humans must take care of it, not for its own sake, but because god ordered humans to do so. Key values are loyalty, humility, sacrifice and respect for authority and tradition.

The modern value system is *world-centric* and holds a cosmopolitan outlook. It came onto the scene in a collectively significant way during the Enlightenment, the modern political revolutions in England, France and the United States and the scientific revolution in Europe roughly spanning the 16th to the 18th centuries. People find their identity as free and equal human beings, first and foremost, regardless of religion, ethnicity, sex, ability or nationality. Moderns tend to see the world as a free market of individuals using their talents and labour to compete and engage in exchange for mutual benefit to attain the most optimal distribution of resources. They tend to be materialistic rather than moralistic: moderns do not place as much emphasis on character formation as do traditionalists. What matters is that one should not physically harm others or take their property. One should respect their rights. The modern value system is thus the father of classical liberalism, the political DNA of modern societies. Key values are freedom, independence, adaptability, risk-taking and experimentation.

The postmodern sensibility came on the scene in the countercultural revolutions of the 1960s and 1970s. It has a deep distrust of any kind of centralized authority, institutional hierarchy, or mainstream establishment. It tends to see these as power structures created to oppress people. It seeks community as a reaction to the alienation brought on by modern individualism and has a deep yearning for existential meaning as a reaction to modern materialism. Postmodernity sees modernity as an artificial environment that has removed us from – and despoiled – nature. It is suspicious of grand narratives about history, which it thinks conceal ideological projects such as colonialism, patriarchy and white supremacy that have historically oppressed and exploited women and people of colour. It goes beyond modern in extending the circle of moral concern to non-human animals and the environment. It thinks from a global and ecological perspective. Key values are tolerance, inclusivity, pluralism, equality and sensitivity.

I suggest that we see these value systems not just as 'different' but, in a developmental context, that they track aspects of cognitive and moral development. An important dynamic in this developmental approach is that each stage attacks and disowns the one before it. In short, we get 'culture wars'. Thus, modern attacks traditional mythology: modernity, the Enlightenment, science debunking religion and so on. Postmodern attacks modern: science, industrialization, reason, capitalism and modern technology dissociate humans from nature and destroy nature. In turn, traditional bemoans modern materialism and postmodern relativism; modern regards postmodern as an enemy of progress; and postmodern rolls its eyes at traditional's denial of science and its religiously founded anthropocentrism.

What is needed, I think, is to acknowledge the relative legitimacy of all of these perspectives, including them but transcending them by recognizing all as elements of our being; they are not types *of* people, but types *in* people. They are developmental capacities at our disposal – the goal is to not be 'stuck' inside any one of them, but to develop a greater identity in which we can activate and inhabit them when appropriate – and they all serve vital purposes. Indeed, each developed as a creative response to emergent life conditions. A few thousand years ago, for instance, traditionalism was a great leap forward for humankind, but over time it can cease to help and instead hinder humanity's ongoing flourishing. The question is not which is 'right' and which is 'wrong', but which is *most right* for a given culture, at a given time, in given circumstances (in relative terms), and which is *more right in the long run* (in absolute terms). That is, in the long run, postmodern is *better* than modern, which is *better* than traditional. What this vision suggests, then, is not just a moral pluralism, but a *developmental* pluralism.

Crudely put, traditionalists deny the problem of climate change, moderns downplay the problem because they believe that the free market and technology will solve it, and postmoderns think it is the greatest problem ever, and that we need to implement major regulations and radical changes in the organization of society and our everyday lives. To decrease the gridlock between these tectonic cultural plates, I suggest that *appeals to climate justice must be both more traditional and less traditional, more modern and less modern, and more postmodern and less postmodern.* Steve McIntosh explains that according to this logic,

> *American culture could become . . . more in touch with the wisdom of our heritage yet less ethnocentric and xenophobic. . . wealthier and better connected yet less concerned with the acquisition of wealth and materialistic possessions. . . more determined to work for social justice and protect the environment yet less hostile toward America's history and its role as the world's leading modernist nation.* (McIntosh 2015: 41)

Appeals to the rights of future generations offered by moderns and postmoderns on cosmopolitan grounds could be framed in traditionalist terms by appealing to the health and flourishing of future citizens who can preserve local and national cultures. Appeals to make the government invest heavily in research and development into decarbonization technologies, which modernist economic conservatives might see as unjust interference in the economy and confiscation of taxpayer money, can be framed as essential to both building a prosperous and globally competitive economy and minimizing the national cost of climate damages. Appeals to policies such as carbon taxes can be framed in terms of climate justice by pointing out that polluters are effectively imposing costs on governments, consumers and taxpayers. Finally, appeals to the rights of nature and non-human species, which typically animate postmoderns, could be framed by linking environmental protection to stewardship for creation and to the tradition of wilderness preservation so central to the American national identity, which would appeal to those with a traditional worldview.

To illustrate the tensions between these worldviews, and the difficulty of integrating them in relation to climate change, I offer a brief analysis of one of the most widely circulated environmental statements of the last few years: Pope Francis' encyclical *Laudato Si'*.

Applying the frameworks: the case of *Laudato Si'*

Environmentalism has taken root primarily in the postmodern worldview but is starting to make inroads among moderns with the rise of 'sustainable' business, and even in traditionalism, many Christians, especially younger evangelicals and, notably, Pope Francis, are beginning to recognize a responsibility to deal with climate change out of a duty to be stewards of God's creation. Indeed, in the encyclical, *Laudato Si': On Care for Our Common Home*, we find the pope scathingly criticizing the modern worldview: the materialism of consumer culture, the greed that drives global capitalism, the rapacious destruction of nature at the hands of what he calls the 'technocratic paradigm' (Bergoglio 2015: 38). He criticizes the exploitation of the Global South by the Global North and recognizes the entanglement of environmental and social justice. He provides a scriptural basis for why Christianity is not, contra the claims of early environmentalists, inherently anthropocentric. I think he threads a needle by arguing that all living things have an intrinsic value (a postmodern characterization) yet insisting that humans take priority (a traditional characterization). Anthony Mills explains how Francis cuts through a dichotomy between 'the amoral language of libertarian technocracy, which sees in humankind the solution to all problems [a modern characterization], and the morally infused and often pantheistic language of environmentalism, which sees humanity as the root of all problems [a postmodern characterization]' (Mills 2015: 54). In other words, Francis critiques the excesses of moderns and postmoderns. Moreover, his 'integral ecology' reframes environmentalism, insisting that we must see it in the context not only of outer ecology – in relation to social and cultural systems – but of inner ecology – our own moral and spiritual development. Finally, perhaps the most important achievement of *Laudato Si'* is that it harbours the potential for a tectonic shift in climate attitudes given that it is pitched to resonate with those people who have a traditional worldview.

Conclusion

This chapter has been less concerned with defending a specific conception of climate justice and more focused on sketching pragmatic pathways for working toward what I venture to call the 'cosmopolitan consensus'. As mentioned above, most ethical analyses of climate change conclude that the rich countries ought to be doing far more than they currently are committed to do under the Paris Agreement to reduce their emissions and invest in adaptation efforts in developing countries. However, such ethical analyses have not yet succeeded in adequately penetrating global and national climate policy and politics. I have argued that, especially in the case of the United States, this is due in part to differences in worldviews and value systems. From this perspective, the climate crisis is not merely or even primarily a crisis of resources, of carbon dioxide concentrations, or of species extinction. It is primarily a crisis of perspectives that lead to different positions on what climate justice looks like. What is needed to advance the 'cosmopolitan consensus' is a more effective 'delivery system' that scholars and practitioners in ethics, law, international relations and public policy can use to creatively frame their arguments and appeals. We need interdisciplinary teams of researchers from philosophy, psychology, sociology, anthropology and political science to synthesize and determine the normative implications of people's values and worldviews for their attitudes toward climate change. This body of research could help to create a 'Rosetta stone' capable of translating catalytic climate policies into the language of every worldview. This could serve as a tool for activists, politicians, policy makers, educators and church leaders to help size up the values and worldview orientations of the communities they work with, and to better educate and motivate them to work for climate justice.

References

Bergoglio, Jorge Mario [Pope Francis] (2015), *Laudato Si': On Care for Our Common Home*, Huntington, IN: Our Sunday Visitor Publishing.

Broome, John (2012), *Climate Matters: Ethics in a Warming World*, New York: W.W. Norton and Co.

Brown, Donald (2013), *Climate Change Ethics: Navigating the Perfect Moral Storm*, New York: Routledge.

Caney, Simon (2010), 'Cosmopolitan justice, responsibility, and global climate change', in Stephen Gardiner et al. (eds), *Climate Ethics: Essential Readings*, New York: Oxford University Press, pp. 122–45.

Dickinson, Janis L., Poppy McLeod, Robert Bloomfield and Shorna Allred (2016), 'Which moral foundations predict willingness to make lifestyle changes to avert climate change in the USA?', *PLoS ONE*, 11(10).

Douglas, Mary (1970), *Natural Symbols: Explorations in Cosmology*, New York: Pantheon Books.

Douglas, M. and Wildavsky, A. B. (1982), *Risk and Culture: An Essay on the Selection of Technical and Environmental Dangers*, Berkeley: University of California Press.

Fritsch, Matthias (2012), 'Climate change justice', *Philosophy and Public Affairs*, 40(3), 225–53.

Frodeman, Robert (2006), 'The policy turn in environmental philosophy', *Environmental Ethics*, 28(1), 3–20.

Gardiner, Stephen M. (2011), *A Perfect Moral Storm: The Ethical Tragedy of Climate Change*, New York: Oxford University Press.

Gardiner, Stephen M. and David Weisbach (2016), *Debating Climate Ethics*, New York: Oxford University Press.

Haidt, Jonathan (2009), *The Righteous Mind: Why Good People are Divided by Politics and Religion*, New York: Vintage Books.

Harris, Paul G. (ed.) (2011), *Ethics and Global Environmental Policy: Cosmopolitan Conceptions of Climate Change*, Cheltenham, UK and Northampton, MA, USA: Edward Elgar Publishing.

Harris, Paul G. (2016), *Global Ethics and Climate Change*, Edinburgh: Edinburgh University Press.

Holling, C.S. (1986), 'The resilience of terrestrial ecosystems: local surprise and global change', in W.C. Clarke and R.E. Mann (eds), *Sustainable Development of the Biosphere*, Cambridge: Cambridge University Press, pp. 217–32.

Hulme, Mike (2009), *Why We Disagree About Climate Change: Understanding Controversy, Inaction, Opportunity*, New York: Cambridge University Press.

Inglehart, Ronald (2018), *Cultural Evolution: People's Motivations are Changing and Reshaping the World*, New York: Cambridge University Press.

Kahan, Dan, Ellen Peters, Maggie Wittlin, Paul Slovic, Lisa Larrimore Ouellette, Donald Braman and Gregory Mandel (2012), 'The polarizing impact of science literacy and numeracy on perceived climate change risks', *Nature Climate Change*, 2, 732–35.

McIntosh, Steve (2015), *Presence of the Infinite: The Spiritual Experience of Beauty, Truth, and Goodness*, Wheaton, IL: Quest Books.

Mills, M. Anthony (2015), 'Is Pope Francis anti-modern?', *New Atlantis*, 47, 45–55.

Moellendorf, Darrel (2015), 'Climate change justice', *Philosophy Compass*, 10(3), 173–86.

Posner, Eric (2013), 'You can have either climate justice or a climate treaty, not both', *Slate*, accessed

19 February 2019 at https://slate.com/news-and-politics/2013/11/climate-justice-or-a-climate-treaty-you-cant-have-both.html.

Sauer, Hanno (2015), 'Can't we all disagree more constructively? Moral foundations, moral reasoning, and political disagreement', *Neuroethics*, **8**, 153–69.

Shue, Henry (2010), 'Subsistence emissions and luxury emissions', in Stephen Gardiner et al. (eds), *Climate Ethics: Essential Readings*, New York: Oxford University Press, pp. 200–214.

Singer, Peter (2010), 'One atmosphere', in Stephen Gardiner et al. (eds), *Climate Ethics: Essential Readings*, New York: Oxford University Press, pp. 181–99.

Stern, Nicholas (2010), 'The economics of global warming', in Stephen Gardiner et al. (eds), *Climate Ethics: Essential Readings*, New York: Oxford University Press, pp. 39–76.

Thompson, Michael (1998), 'Cultural discourses', in S. Rayner and E.L. Malone (eds), *Human Choice and Climate Change*, Columbus, OH: Batelle Press, pp. 265–344.

Vainio, Annukka and Jaana-Piia Mäkiniemi (2016), 'How are moral foundations associated with climate-friendly consumption?', *Journal of Agricultural and Environmental Ethics*, **29**, 265–83.

World Values Survey (2014), accessed 17 February 2019 at www.worldvaluessurvey.org/.

5   National climate-mitigation policy: the spatial framing of (in)justice claims

Ian Bailey

Mary Robinson, founder of the Mary Robinson Foundation: Climate Justice, describes climate justice as a human-centred approach to managing climate change based around safeguarding the rights of people most affected by climate change while ensuring that justice also underpins low-carbon transitions, including people's right to development (Robinson 2018). These are compelling ideals, yet progress towards integrating climate justice into policy remains faltering in many countries, particularly when it comes to measures to reduce greenhouse-gas emissions. Controversies over climate mitigation policy typically centre on the economic impacts of initiatives and their implications for consumer lifestyles, but it is also important to appreciate the potential for actions to control climate change to trigger disputes over how notions of fairness should be defined and applied where climate policies impinge on other societal justice concerns (Bailey 2017).

My goal in this chapter is to argue that climate justice scholarship and activism need to pay greater attention to how justice on climate issues is contested within real-world political debates. The climate justice literature has made important progress in articulating the principles of climate justice and in pressing for their incorporation into decision-making (Okereke 2010, Schlosberg 2012). However, less attention has been paid to how climate justice negotiates its relationship with other economic and social justice concerns affected by climate initiatives. My contention is that closer examination of how notions of justice are debated within climate politics is essential to avoid climate justice being treated as somehow isolable from, or axiomatically superior to, other justice concerns. What is needed instead is a consciously relational outlook in climate justice research that actively explores how tensions between climate justice and other forms of justice are expressed and influence the development of climate policy (Klinsky et al. 2012, Schlosberg et al. 2019). This also implies a need for conceptual analysis to be complemented to a greater degree by empirical investigation of the distinctive politics, power relations and opportunities influencing how climate justice is defined, debated and applied in different geographical settings (Bulkeley et al. 2014).

Thinking about justice on climate issues as a contested concept in turn invites attention to the phenomenon of *justice claim-making* in climate politics, whereby actors participating in policy debates employ justice arguments to defend standpoints that might not gain support unless attached to an (in)justice claim (Beckman and Page 2008). In simple terms, disagreeing with a proposed action because of differences of opinion or because it clashes with an actor's interests is likely to be less persuasive than if the action can be portrayed as unjust. Consideration then needs to be paid to the tactics used to gain traction for such (in)justice claims. These may include mobilizing accepted discourses about the importance of economic growth for societal well-being, utilizing the media to promote viewpoints (Boykoff and Boykoff 2007), or personalizing issues by identifying victims and perpetrators of climate and other injustices (DiFrancesco and Young 2011). Another less researched tactic involves attaching spatial representations to (in)justice claims in order to enhance their appeal with key audiences by creating links between injustices and identifiable groups and/or places (Bailey 2017). Corporations, for instance, might be given short shrift for stressing the commercial effects of climate policies, even if they accuse governments of acting prejudicially against the company or sector. However, similar arguments may gain greater sympathy if corporations stress the risks to national economies or employment in areas that depend on the sector. Alternatively, opponents of climate initiatives may attempt to influence policy by accusing governments of imposing unfair and irresponsible burdens on the economy because other countries are not taking on similar responsibilities to reduce their emissions. In essence, these and other types of spatial anchoring seek to appeal to the concerns of selected audiences by crafting tailored messages about the injustices of climate action or inaction, and can particularly influence policy debates where they also target politicians' electoral incentives and media reporting further erodes public support for climate action.

Spatial framing and justice claim-making in national climate policy

To explore the influence of justice claim-making on climate policy, I examine the ways spatial constructions of (in)justice have shaped the evolution of national climate mitigation policy in Australia, New Zealand, the United States and the United Kingdom. These countries provide interesting contrasts in the use of spatial representation to depict justice issues associated with climate change and climate policy. Climate change remains a politically venomous issue in Australia and the US, and even when they have tried, the two federal governments have made limited headway in establishing coherent national mitigation strategies. In the US, the Trump administration has sought to repeal many of the climate measures introduced by Donald Trump's predecessor, Barack Obama, while Australia has engaged in protracted and acrimonious debates over the introduction of carbon pricing. A Carbon Pricing Mechanism was legislated by Julia Gillard's minority Labor government in 2011 after two previous failed attempts, but this was repealed in 2014 by the successor centre-right coalition administration, led by Tony Abbott (Bailey 2017). In New Zealand, a national emissions trading scheme started operating in late 2008. However, until the election of a Labour-led coalition led by Jacinda Ardern in 2017, the policy was constructed in ways that avoided exerting significant downward pressure on emissions (Bertram and Terry 2010). In the

UK, the Climate Change Act of 2008 established a statutory long-term emissions target to reduce greenhouse-gas emissions to 80 per cent below 1990 levels by 2050 and five-year interim carbon budgets. However, the policy has moved into more challenging terrain since 2010 as deeper emissions cuts have been required and tensions have accumulated within the UK's cross-party consensus on climate change (Gillard 2016).

The evidence from these countries indicates that opponents of stronger climate action have shown considerable skill in developing spatially and socially recognizable discourses about the injustices of climate policy, emphasizing the failure of other countries to act and the economic impacts of policies on vulnerable regions. Supporters of stronger climate mitigation policy, in contrast, have relied more heavily on narratives highlighting the responsibility on wealthier nations to lead low-emissions transitions. National and local climate justice issues have also featured in pro-climate action narratives but usually in abstract and long-term future-gazing ways that have weakened their traction in debates on national climate policy. Particularly within the US, Australia and New Zealand, clear associations can be identified between the use of national, regional and local injustice claims to obstruct or dilute climate initiatives and the difficulties experienced by governments in introducing or strengthening national climate mitigation policy.

Framing climate justice in an international and national context

The portrayal of climate change as an international justice issue has been one of the main arguments used by non-government organizations and academics to press for stronger climate action by wealthier nations. In its more straightforward usage, it invokes a non-judgemental moral responsibility to protect fellow human beings around the world and in the future from the adverse impacts of climate change (Gardiner 2011). Other variants are more forthright about the need to redress historical and ongoing injustices experienced by countries in the Global South resulting from the ecologically unequal exchange of resources and energy with the Global North through the application of international climate justice principles (Godard 2017, Roberts and Parks 2009).

Such global representations of climate justice have also been employed extensively by political leaders to build support for climate action internationally and in their own countries. In their speeches at the United Nations Climate Change Conference in Paris in 2015, US President Obama and Chinese Premier Xi Jinping both stressed the international justice dimensions of climate change to communicate their agendas to their negotiating partners and domestic audiences. When Obama declared that, 'As the leader of the world's largest economy and the second largest emitter ... the United States of America not only recognizes our role in creating this problem, we embrace our responsibility to do something about it', he used international justice framings simultaneously to acknowledge the US's responsibilities and to press for reciprocal commitments from other countries (Obama 2015). This framing, also aimed at convincing political actors at home that climate action by the US would not threaten its economy, was mirrored in Xi Jinping's call for international partnership and equity, a reassurance to other countries about China's intentions but equal insistence that any deal should enable China to continue its economic and development strategies: 'Tackling climate change is a shared mission for mankind . . . Let us join hands to contribute to the establishment of an equitable and effective global mechanism on climate change, work for global sustainable development at a high level and bring about new international relations featuring win–win cooperation' (Xi 2015).

While politicians have used international climate justice framings both for offertory purposes and to assert entitlements in international negotiations, they have also been used extensively in domestic politics to legitimate new climate initiatives. One more notorious example of this was former Australian Prime Minister Kevin Rudd's repeated description of climate change as a 'great moral challenge' between 2007 and 2010 during debates on his government's proposed Carbon Pollution Reduction Scheme (Rudd 2008). Although the scheme faced strong industry opposition and was withdrawn after being rejected twice by the Australian Senate (Bailey et al. 2012), other leaders have enjoyed greater success using international framings of climate justice. As UK Prime Minister, Tony Blair was energetic in using moral framings to push for international cooperation on climate change and as a way of laying the ground for the UK's Climate Change Act. In 2004, Blair announced his desire for major industrialized nations to show greater climate leadership, declaring that: 'The world's richest nations in the G8 have a responsibility to lead the way: for the strong nations to better help the weak' (Blair 2004).

Accentuating the international climate responsibilities of wealthier countries, while ethically and practically self-evident to some, nevertheless exposes new climate measures to accusations of proffering one-sided sacrifices. One option to restrict the scope for counterclaims of national injustice is to develop messages aligning proposed actions with core national interests. This reasoning featured prominently in the 2008 and 2011 Garnaut climate change reviews commissioned to inform Australia's Carbon Pollution Reduction Scheme and Carbon Pricing Mechanism, both of which claimed that climate action by Australia offered national benefits in combating climate insecurity and would give the country competitive advantages in low-carbon innovation (Garnaut 2008, 2011). The corresponding sub-text was that failing to act would lead to adverse consequences and, by extension, injustice for future generations. According to the 2011 Garnaut review,

> Modelling showed that the growth rate for Australian national income in the second half of the 21st century would be higher with mitigation than without. The present value of the market benefits this century fell just short of the costs of mitigation policy. However, when we took account of the value of Australians' lives beyond the 21st century, the value of our natural and social heritage, health and other things that weren't measured in the economic modelling, and the value of insuring against calamitous change, strong mitigation was clearly in the national interest. (Garnaut 2011: x)

One conspicuous omission from these portrayals, however, was any direct invocations of justice to underpin the case for climate action. National interest was instead presented as a matter of rationality, supported by general referents to social heritage, health and 'national utility or welfare over time' (Garnaut 2008: 15). It is perhaps not difficult to see how, for public and business audiences contemplating the prospect of paying more for goods and services as a result of carbon pricing, the lack of reinforcement of links between climate change and fairness left them unclear as to what inaction on climate change meant for them and unpersuaded of the value of carbon pricing except in respect of altruism for future generations.

Opponents of climate mitigation often offer starkly different interpretations of international climate justice to rationalize their stance against the strengthening of measures to reduce greenhouse gas emissions. One common tactic is to contrast proposed measures by one country with the absence of equivalent action by other countries as a way of arguing that the policy would create disproportionate risks to critical national interests and should be recalibrated to reflect commitments made by other countries (Garnaut 2011). From the George W. Bush administration to the Trump administration in the US, officials have routinely used national competitiveness and free-rider arguments to depict China and India as gaining unfair advantages in the global economy at the expense of US interests (Harris 2013, 2016). Defensive comparisons with other countries have also been a recurring theme in Australian climate politics, from John Howard's refusal to ratify the Kyoto Protocol in 2003 for fear it 'would destroy jobs and the competitiveness of Australian industry' to Tony Abbott's pitch to make the 2014 general election a referendum on Australia's 'economically devastating' carbon tax (Rootes 2014, *The Age* 2003).

This reasoning has proven especially effective in stirring up anti-climate justice sentiments when used in conjunction with narratives emphasizing the insignificance of actions by individual countries in reducing global emissions. Perhaps unsurprisingly, insignificance has formed a key argument among actors resisting reforms to strengthen the New Zealand emissions trading scheme (Bailey and Jackson Inderberg 2016). The New Zealand Labour Party has challenged this assessment, arguing: 'It is not good enough to say we are too small to matter – most countries individually could claim the same' (Labour 2017). Aligning sectoral interests with national interests has nevertheless formed a key defence against pricing biological emissions from agriculture, despite the sector making up nearly 50 per cent of New Zealand national emissions. The foundation for this argument has been that agriculture is vital to New Zealand's national economy and that pricing biological emissions would damage the sector's competitiveness, unless other countries introduced similar measures, because it is strongly export-oriented and a price-taker on international commodity markets. Not only would New Zealand farmers be penalized, the argument continues, the move would yield no reduction in global emissions because production would simply move to countries with less emissions-efficient agricultural sectors (Cooper and Rosin 2014). Prior to winning the Australian Prime Ministership, Tony Abbott similarly argued that Australia only accounted for one per cent of global carbon emissions and that China's annual increase in emissions matched Australia's entire carbon output. According to this logic, climate leadership by Australia risked damaging industry but would make limited difference to global emissions (Abbott 2009), and even the European Union felt the need to introduce safeguards against carbon leakage arguments when reforming its emissions trading scheme in 2009 (van Asselt and Brewer 2010). However, the extra strength of justice arguments against pricing agricultural emissions in New Zealand was its focus on a profession respected for its industry and (sometimes flatteringly) for its custodianship of New Zealand's green, clean image (Cooper and Rosin 2014), a tactic that magnified the sense of injustice against New Zealand's national interests from the implementation of 'reckless' climate measures.

Even in countries with more ambitious climate policies, comparable justice claims and counterclaims have occurred. In the UK, for example, Carter and Jacobs (2014) argue that the introduction of the 2008 Climate Change Act was significantly aided by the government's use of three focusing events – the Gleneagles G8 Summit in 2005, the release of *An Inconvenient Truth* in 2006, and the publication the *Stern Review* (Stern 2007) – to frame climate change as an urgent environmental and economic issue. In particular, the economic frame – based around the threats to the global and UK economy from climate change and the national rewards from building a low-carbon economy – combined with Prime Minister Blair's moral oratory helped to build new constituencies for climate policy and enlarged the political space for policy development. However, later institutional analysis by Gillard (2016) suggests the surfacing of more inward-looking portrayals of the UK's national interest following the global financial crisis and the onset of government austerity. This shift was expressed in a diminishing accent on the UK's moral imperative to act urgently against dangerous climate impacts and to correct previous policy failures in favour of rhetoric justifying the setting of future carbon budgets based on their ability to produce economic and social co-benefits. The practical effect of this included the replacement of support across the political spectrum for carbon budgets that were as ambitious as economically possible during the first three budget-setting rounds with a more cautious and contested approach to the fourth and fifth budgets (Gillard 2016).

Contesting climate justice at the regional and local level

Supporters and opponents of climate action have both framed climate change as an international and national justice issue. However, opponents of new climate initiatives have been conspicuously more energetic in invoking claims about the regional and local injustices of climate mitigation policy. During the 2016 US presidential election campaign, Donald Trump made major gains in states like Wyoming, West Virginia, Kentucky and Pennsylvania by pledging to revive the fortunes of coal mining and other energy-intensive industry communities: 'We will unleash America's energy, including shale, oil, natural gas and clean coal. We will put our miners back to work. We will put our steel workers back to work' (Schrock et al. 2017: 14). The critical feature of Trump's campaign was not simply its pinpointing of justice narratives towards regions where there was a chance of electoral gains, but also the appeal to wider emotional grievances about overregulation and the neglect of US regions and blue-collar workers by Washington elites, while at the same time rebuking previous administrations for allowing other countries to steal unfair advantages. Trump also found an unintentional ally on this issue in Hillary Clinton, who, at a town hall meeting in Columbus, Ohio, in March 2016, remarked that she would bring economic opportunity in coal-mining areas through renewable energy and 'put[ting] a lot of coal miners and coal companies out of business' (Clinton 2017). Although Clinton's intention was to project a vision of positive transitions for coal-reliant regions and the global climate, it drew fierce criticism for appearing to trivialize coal workers' concerns and deny them fair treatment.

Regional and local injustice claims similarly became a hallmark of debates over Australia's Carbon Pricing Mechanism. Barnaby Joyce, the then Shadow Minister for Regional Australia, Regional Development and Local Government, and an outspoken critic of carbon pricing, orchestrated media releases specifying areas he claimed would be devastated by a measure the opposition political parties pejoratively dubbed 'the carbon tax':

> New South Wales [NSW] Treasury figures show that the carbon tax will lead to 31,000 lost jobs in NSW but over 26,000 of these would be in regional Australia, including 18,500 in the Hunter, 7,000 in the Illawarra and 1,000 jobs in the central West. . .. If [the minister] wants to continue his 'embrace the challenge' tour of regional Australia he needs to come clean [about] how they are meant to embrace the challenge of fewer employment opportunities. . .. A carbon tax will clearly hurt regional Australia the worst. (Joyce 2011)

Parallel warnings from *The Australian* newspaper swelled this regional injustice narrative:

> Explosive economic modelling warns that the carbon tax could force eight black [bituminous] coalmines to close, costing nearly 3000 jobs in regional NSW and more than 1100 jobs in Queensland in its first three years. Independent modelling commissioned by the Australian Coal Association warns that the number of early mine closures could reach 18 within nine years and result in Australia forgoing coal sales of $22 billion from existing mines over the next decade. (*The Australian* 2011)

Other noteworthy features of these portrayals were their combining of regional injustice narratives with references to 'independent' or government modelling and the specification of short- and longer-term effects to authenticate claims about carbon pricing while heightening their cognitive-emotive impact through rhetoric seeking to trigger aggrieved reactions (Schrock et al. 2017). This linguistic tactic was extended to the level of individual cities and households through warnings of blackouts as energy companies struggled with the costs of the pricing mechanism, Tony Abbott's references as leader of the opposition to carbon pricing as 'a great big tax on everything', and accusations that the price of a Sunday roast dinner would spiral to AU$100 (McNair 2014). Such claims further personalized and dramatized carbon pricing by constructing innocent victims and by suggesting the policy would negatively affect every aspect of people's lives. Outrage about these 'instantaneous' impacts was further heightened by scientifically questionable allegations about the lack of environmental benefits of carbon pricing. In 2012, the year following the introduction of the Carbon Pricing Mechanism, Joyce – a renowned political brawler – remarked during an interview: 'Has it become remarkably colder? Are we now living in a global nirvana because we've brought in the carbon tax? No, it's exactly where we left it. However, people are definitely poorer because of the carbon tax, and it's done nothing to the climate' (*Sydney Morning Herald* 2012).

The portrayal of regional victims in injustice arguments also featured heavily in disputes over whether to include agricultural biological emissions in the New Zealand emissions trading scheme. As noted earlier, the imagery of New Zealand as an agricultural nation continues to hold high currency in New Zealand society despite growing concerns about the impacts of agricultural intensification on forests, soil erosion and water quality (Driver et al. 2018). The prospect of pricing biological agricultural emissions consequently sparked accusations of farmers being persecuted by 'city-dwelling greenies' who misunderstood the short lifespan of methane emissions and the lack of opportunities available for farmers to reduce biological emissions except for destocking and converting productive land to forestry, both of which eroded rural livelihoods (Jackson Inderberg et al. 2018). Further disagreements emerged over proposals for sector-level monitoring and enforcement of agricultural emissions, which lobby groups claimed denied individual farmers the opportunity to benefit from cost-effective farm-scale initiatives (Cooper and Rosin 2014). Framing this as a victims-and-villains story in which well-meaning farmers have been prevented from 'doing the right thing' by urban elites has strong similarities with the divisive international comparisons used by US and Australian governments to justify refusal to adopt stricter emissions targets and policies. It also reinforces the importance of identifying worthy, downtrodden victims to contrast against blameworthy perpetrators to accentuate the perceived injustice of new climate policies (Schlosberg 2012). In reality, New Zealand agriculture is a well-organized, industrialized sector that benefits from significant foreign direct investment, but cultural images of farmers as guardians of Arcadian ideals and bedrocks of the economy have fuelled emotive claims about lack of fairness in the design of emissions trading in New Zealand (Cooper and Rosin 2014).

Importantly, discussion of the regional and local injustices of climate policy has been noticeably more muted in countries that have made greater progress in developing national mitigation strategies. The UK especially illustrates the effects of consensus among the main political parties in diminishing the capacity of climate policy opponents to provoke controversy through injustice claims. In contrast, sectarian climate politics and the greater lobbying power of extractive and energy-intensive industries in the US and Australia have contributed to rancorous disputes over the justice implications of climate policy, where attacks have continued to haemorrhage the credibility of policies even after their adoption. New Zealand represents a more mixed case; although the National Party resisted emissions trading legislation while in opposition, neither main political party has been prepared to oppose carbon pricing outright and debates have instead focused on how much action New Zealand should take to do its 'fair share' on emissions reduction – to avoid falling foul of global justice narratives – while not undermining the economy and well-being of key sectors. Once in government, the National Party nevertheless used national and local justice arguments to justify reducing the scheme's practical impact. The spatial framing of (in)justice claims, either for ethical reasons or pre-textually to defend vested interests, has nevertheless exerted a profound influence on climate policy across the countries examined.

Rethinking the spatial representation of climate justice

Particularly in the US, Australia and New Zealand, opponents of stronger action on climate change have scored notable successes in obstructing or diluting climate initiatives by employing multi-scalar injustice narratives that stress the unfair impacts of measures like carbon pricing at the national, regional, local and household levels. In contrast, advocates of stronger climate action have tended to rely more heavily on representations of climate change as a global justice issue and on appeals for wealthier nations to show climate leadership. Even where they have attempted to stress the long-term national benefits of action, these have rarely included explicit messaging about the potential distributive injustices for their countries of failing to act on climate change. The evidence indicates that these more abstract representations of climate justice often struggle to capture the imagination of audiences compared with the more personalized justice messaging of their political rivals.

If climate justice is to become more of a transformative force in national climate politics, climate justice scholars and practitioners need to find more imaginative ways to spatialize the justice arguments for stronger climate action. This is likely to require a more direct focus on national, regional and local concerns, and on co-benefits over and above stand-alone arguments about constraining and managing climate change. Hillary Clinton herself expressed regret at saying she would put coal miners and coal companies out of business even though she also stressed the prospect of bringing new economic opportunities to coal country by incentivizing renewable energy (Clinton 2017). The negative element of this message dominated media coverage when an alternative framing reminding audiences of the harmful effects of the coal industry on the health and well-being of people living in coal-mining areas may have made them more responsive to her vision of a clean energy future. New ways of spatializing the justice arguments for greater climate action might equally invoke images of the possible effects of climate change on individual areas and sections of society, though the spatial and temporal unpredictability of climate impacts makes such framings prone to allegations of alarmism and of making misleading connections between weather events and climate change. Equally, challenges exist in finding framings that increase the *saliency* of climate justice while also offering persuasive narratives about the possibility of effective and equitable action (O'Neill et al. 2013). The potential for similar spatial framings to produce different responses in different countries also underscores the need for bespoke approaches to reduce psychological distancing between audiences' underlying sympathies with the notion of climate justice and the practicalities of advancing climate policy (Spence et al. 2012).

Possibilities for navigating such difficulties include aspirational narratives emphasizing the potential for cities and regions to become hotspots of cleantech innovation, and governments as facilitators of effective and fair transitions from carbon-intensive activities (Gibbs and O'Neill 2014). More premonitory framings might stress the competitive risks and unequal outcomes of lagging behind in the de-carbonization of the global economy, although research indicates that caution-ary messages often elicit mixed responses, while unfocused warnings may again lack credibility (Spence et al. 2012). Either way, increasing the relevance of the justice issues associated with climate change to the locality of audiences and identi-fiable social groups is crucial in promoting greater willingness to accept or tolerate new mitigation responsibilities.

Conclusion

This chapter has argued that research on climate justice needs to pay greater attention to exploring conflicts between climate justice and other societal justice concerns. Climate justice scholarship has made important strides in exploring the principles of a distinctive climate justice and how these ideas should inform decision-making on climate issues. However, it has paid less attention to how ten-sions between climate justice and other justice concerns have hampered attempts by governments to respond to the challenges of mitigating and adapting to climate change (Barrett 2012, Bulkeley et al. 2014). Understanding how these tensions are expressed and managed requires scholars to move beyond normatively focused analyses of climate justice and to develop deeper understandings of how political actors utilize arguments about the economic and social injustices of climate action (or inaction) to build legitimacy and support for their stances towards climate initiatives. Examining climate justice instead as one of many interpretations of justice competing to influence climate policy may provide important insights on why – beyond national and commercial interests – rights-centred approaches to managing climate change often struggle to achieve political traction.

Spatial (in)justice framings have been used extensively within national climate poli-tics to accentuate and give personality to justice arguments for and against more purposeful action on climate change. So far, opponents of stronger climate policy have spatialized the injustices of climate initiatives in more diverse and persuasive ways than have those advocating more decisive action. Regardless of whether these framings reflect ulterior motives or sincere concerns about the impacts of climate policies on national, regional and local economies, climate justice research needs to respond by finding new and imaginative ways to spatialize the justice arguments for stronger climate action.

References

Abbott, Tony (2009), *Battlelines*, Carlton: Melbourne University Press.

Bailey, Ian (2017), 'Spatializing climate justice: justice claim making and carbon pricing controversies in Australia', *Annals of the American Association of Geographers*, **107**(5), 1128–43.

Bailey, Ian and Tor Håkon Jackson Inderberg (2016), 'New Zealand and climate change: what are the stakes and what can New Zealand do?', *Policy Quarterly*, **12**(2), 3–12.

Bailey, Ian, Iain MacGill, Robert Passey and Hugh Compston (2012), 'The fall (and rise) of carbon pricing in Australia: a political strategy analysis of the Carbon Pollution Reduction Scheme', *Environmental Politics*, **31**(5), 691–711.

Barrett, Sam (2012), 'The necessity of a multiscalar analysis of climate justice', *Progress in Human Geography*, **37**(2), 215–33.

Beckman, Ludvig and Edward Page (2008), 'Perspectives on justice, democracy and global climate change', *Environmental Politics*, **17**(4), 527–35.

Bertram, Geoff and Simon Terry (2010), *The Carbon Challenge: New Zealand's Emissions Trading Scheme*, Wellington: Bridget Williams Books.

Blair, Tony (2004), 'Full text: Blair's climate change speech', *Guardian.com*, Green politics, 15 September, accessed 4 February 2019 at https://www.theguardian.com/politics/2004/sep/15/greenpolitics.uk.

Boykoff, Maxwell T. and Jules M. Boykoff (2007), 'Climate change and journalistic norms: a case-study of US mass-media coverage', *Geoforum*, **38**(6), 1190–204.

Bulkeley, Harriet, Gareth Edwards and Sara Fuller (2014), 'Contesting climate justice in the city: examining politics and practice in urban climate change experiments', *Global Environmental Change*, **25**, 31–40.

Carter, Neil and Michael Jacobs (2014), 'Explaining radical policy change: the case of climate change and energy policy under the British Labour Government 2006–10', *Public Administration*, **92**(1), 125–41.

Clinton, Hillary R. (2017), *What Happened*, New York: Simon and Schuster.

Cooper, Mark H. and Christopher Rosin (2014), 'Absolving the sins of emission: the politics of regulating agricultural greenhouse gas emissions in New Zealand', *Journal of Rural Studies*, **36**, 391–400.

DiFrancesco, Darryn Anne and Nathan Young (2011), 'Seeing climate change: the visual construction of global warming in Canadian national print media', *Cultural Geographies*, **18**(4), 517–36.

Driver, Elizabeth, Meg Parsons and Karen Fisher (2018), 'Technically political: the post-politics(?) of the New Zealand emissions trading scheme', *Geoforum*, **97**, 253–67.

Gardiner, Stephen M. (2011), *A Perfect Moral Storm: The Ethical Tragedy of Climate Change*, Oxford: Oxford University Press.

Garnaut, Ross (2008), *The Garnaut Climate Change Review: Final Report*, Cambridge: Cambridge University Press.

Garnaut, Ross (2011), *The Garnaut Review 2011: Australia in the Global Response to Climate Change*, Cambridge: Cambridge University Press.

Gibbs, David and Kirstie O'Neill (2014), 'The green economy, sustainability transitions and transition regions: a case study of Boston', *Geografiska Annaler B, Human Geography*, **96**(3), 201–216.

Gillard, Ross (2016), 'Unravelling the United Kingdom's climate policy consensus: the power of ideas, discourse and institutions', *Global Environmental Change*, **40**, 26–36.

Godard, Olivier (2017), *Global Climate Justice: Proposals, Arguments and Justification*, Cheltenham, UK and Northampton, MA, USA: Edward Elgar Publishing.

Harris, Paul G. (2013), *What's Wrong with Climate Politics and How to Fix it*, Cambridge, UK and Malden, MA, USA: Polity Press.

Harris, Paul G. (2016), 'Climate change and American foreign policy: an introduction', in Paul G. Harris (ed.), *Climate Change and American Foreign Policy*, New York: Palgrave Macmillan, pp. 3–25.

Jackson Inderberg, Tor Håkon, Ian Bailey and Nichola Harmer (2018), 'Designing New Zealand's emissions trading scheme', *Global Environmental Politics*, **17**(3), 31–50.

Joyce, Barnaby (2011), 'Government must come clean on carbon tax's impact on regional Australia', *Media release*, 23 August 2011, accessed 3 April 2019 at https://barnabyisright.com/category/uncategorized/page/85/?iframe=true&preview=true%2Ffeed%2F.

Klinsky, Sonja, Hadi Dowlatabadi and Timothy McDaniels (2012), 'Comparing public rationales for justice trade-offs in mitigation and adaptation climate policy dilemmas', *Global Environmental Change*, **22**(4), 862–76.

Labour (2017), 'Real action on climate change', accessed 25 January 2019 at https://www.labour.org.nz/climatechange.

McNair, Brian (2014), 'Hard news: the carbon tax shows up cracks in media reporting', *TheConversation.com*, Energy and Environment, 20 July, accessed 3 April 2019 at http://theconversation.com/hard-news-the-carbon-tax-shows-up-cracks-in-media-reporting-29206.

Obama, Barack (2015), 'Remarks by President Obama at the first session of COP21', *UNFCCC.int*, 30 November, accessed 25 January 2019 at https://unfccc.int/sites/default/files/cop21cmp11_leaders_event_usa.pdf.

Okereke, Chukwumerije (2010), 'Climate justice and the international regime', *WIRES: Climate Change*, **1**(3), 462–74.

O'Neill, Saffron, Maxwell T. Boykoff, Simon Niemeyer and Sophie Day (2013), 'On the use of imagery for climate change engagement', *Global Environmental Change*, **23**(2), 413–21.

Roberts, J. Timmons and Bradley Parks (2009), 'Ecologically unequal exchange, ecological debt, and climate justice: the history and implications of three related ideas for a new social movement', *International Journal of Comparative Sociology*, **50**(3–4), 385–409.

Robinson, Mary (2018), 'Climate justice: putting people first', speech to One Young World, The Hague, 20 October, accessed 7 January 2019 at https://www.mrfcj.org/wp-content/uploads/2018/10/Mary-Robinson-Keynote-Speech-OYW-2018.pdf.

Rootes, Christopher (2014), 'A referendum on the carbon tax? The 2013 Australian election, the Greens, and the environment', *Environmental Politics*, **23**(1), 166–73.

Rudd, Kevin (2008), 'Building a better world together', speech at Kyoto University, 9 June, accessed 2 October 2018 at http://pmtranscripts.dpmc.gov.au/release/transcript-15955.

Schlosberg, David (2012), 'Climate justice and capabilities: a framework for adaptation policy', *Ethics and International Affairs*, **26**(4), 445–61.

Schlosberg, David, Karin Bäckstrand and Jonathan Pickering (2019), 'Reconciling ecological and democratic values: recent perspectives on ecological democracy', *Environmental Values*, **28**(1), 1–8.

Schrock, Douglas, Benjamin Dowd-Arrow, Kristen Erichsen, Haley Gentile and Pierce Dignam (2017), 'The emotional politics of making America great again: Trump's working class appeals', *Journal of Working-Class Studies*, **2**(1), 5–22.

Spence, Alexa, Wouter Poortinga and Nick Pidgeon (2012), 'The psychological distance of climate change', *Risk Analysis*, **32**(6), 957–72.

Stern, Nicholas (2007), *The Economics of Climate Change: The Stern Review*, Cambridge: Cambridge University Press.

Sydney Morning Herald (2012), 'Joyce's $100 roast prediction rubbished', accessed 14 February 2019 at https://www.smh.com.au/politics/federal/joyces-100-roast-prediction-rubbished-20121118-29kln.html.

The Age (2003), 'I won't ratify Kyoto Protocol: PM', accessed 10 February 2019 at http://www.theage.com.au/articles/2003/12/02/1070127416251.html.

The Australian (2011), 'Carbon tax will cost 4000 coal jobs', *ABC.net.au*, 14 June, accessed 19 January 2019 at http://www.abc.net.au/mediawatch/transcripts/1119_australian1.pdf.

van Asselt, Harro and Thomas Brewer (2010), 'Addressing competitiveness and leakage concerns in climate policy: an analysis of border adjustment measures in the US and the EU', *Energy Policy*, **38**(1), 42–51.

Xi, Jinping (2015), 'Work together to build a win–win, equitable and balanced governance mechanism on climate change', *UNFCCC.int*, 30 November, accessed 11 February 2019 at https://unfccc.int/sites/default/files/cop21cmp11_leaders_event_china.pdf.

6   Climate change and capitalism: a degrowth agenda for climate justice

Carlos Tornel

The climate justice movement in its three different articulations – as a social movement, as an academic concept and as an elite perspective of environmental non-governmental organizations (ENGOs) (Schlosberg and Collins 2014) – has undoubtedly shaped international agreements and agendas (Okereke and Coventry 2016). For example, the 2015 Paris Agreement on climate change and the United Nations' Sustainable Development Goals (SDGs) were held as a major victory for the climate justice movement at the global level (WWF 2016, WRI 2016). Yet, both the Paris Agreement and the SDGs perpetuate the paradoxical condition where climate change is extensively debated while little actually changes (Swyngedouw 2017). The problem is not the recognition of demands for climate justice, but the actual proposed and dominant solutions to address them.

In this chapter, I argue that the demands of the climate justice movement, which move beyond a traditional approach of rights and responsibilities of nation-states to reduce emissions, have systematically postponed actual climate justice actions in favour of managerial and technology-based responses promoted by political elites, thereby infinitely postponing actions that could lead to real and effective enforcement of such demands. Climate justice has become an 'empty signifier', a discourse that articulates a series of demands subject to 'radically diverse interpretations and, as such, prevents anything concrete from being done' (Brown 2016: 116). This condition presents climate change as a 'post-political populist project' where climate change 'does not invite a transformation of the existing socio-ecological order, but rather calls on the elites to undertake action such that nothing really has to change' (Swyngedouw 2017: 304).

An alternative to this lack of change can be found in *degrowth*. In this chapter, I outline a series of propositions to move towards degrowth as a means of achieving climate justice. I present some of the key proposals for degrowth and a more politically based approach to climate justice. I conclude that climate justice, at least in two of its more prominent manifestations – an academic concept and a slogan for a social movement – must necessarily adopt a critique of capitalism and the post-political condition that emerges from the international climate negotiations and the solutions that have emerged from them. This includes critiquing administrative mechanisms, knowledge structures and regulatory decisions that maintain and perpetuate the exercise of particular ideas and framings of the climate problem.

Conceptualizations of the climate justice movement

The environmental justice movement was originally conceived in specific local contexts with particular uneven exposure to environmental risks and degradation. These concerns were broadly based on the right of local communities not to suffer from an inequitable distribution of environmental risks and governmental protections (Schlosberg and Collins 2014). The movement also became increasingly concerned that the rights of local communities and vulnerable groups (such as indigenous communities and the poor) be considered and recognized when deciding on the development, implementation and outcomes of policies. With respect to global climate change, the justice framework was adopted by a much wider movement. It was seen as a manifestation of a broader social injustice that went beyond a disproportionate burden of already vulnerable communities. Under climate change, vulnerable groups suffer a sort of 'triple injustice' as they are historically and proportionately less responsible for climate change while being disproportionately burdened with the ill effects, lack of recognition and exclusion from political decision-making, and because they could be affected by mitigation or adaptation strategies (Schlosberg and Collins 2014).

At the international level, the issue of climate justice has been mainly a concern of rights and responsibilities between developed and developing countries (Bulkeley et al. 2014). However, there have been calls to surpass the assumption that the nation-state is the only relevant actor – to supplement international climate justice with a cosmopolitan approach whereby responsibility for emissions accrues to people as well as states, and to consider that international negotiations often obscure issues *within* nations (Harris 2010, 2016).

Bulkeley et al. (2014) argue that achieving climate justice must move beyond the traditional notion of balancing rights/responsibilities – that is, who has the right to emit and who has the responsibility to ameliorate the effects of climate change – on one axis, and distribution and procedure on another axis. They present a common denominator to the two axes: that of *recognition*, which must be understood as moving beyond a common responsibility to differentiated approaches and possible injustices created by the climate policy, such as the cultural and symbolic injustices that emerge from distributional approaches to address climate change. Recognition is a useful concept to show both the multidimensional nature of climate justice and the differentiated role of places, specific groups and particular individuals.

This three-dimensional perspective – rights/responsibilities, distribution/procedures and recognition – allows us to understand the roles of different actors and networks that have been persistent throughout the shaping of the international climate change regime. From this perspective, there are three main conceptualizations of climate justice: (1) as a field of the academic literature; (2) as a grassroots global movement; and (3) as an international, elite-based, non-governmental organization approach. The academic approach has been focused mainly on developing a series of concepts, such as the demand for historical responsibility, per-capita equity approaches and rights-based approaches, which have been influential for both the international negotiation process and the global grassroots movement on climate justice (Schlosberg and Collins 2014).

Although there is no globally coordinated climate justice movement led by a single organization, there is a common set of goals, principles and forms of mobilization worldwide that locally advocate for better distribution of the cost and benefits, recognition and representation in the fight against climate change (Martínez-Alier et al. 2016). This movement is both local and global, and it is inherently tied to the changing social metabolism and expansion of the global capitalist economy. That economy is understood as the increasing demand for energy and materials, progressively expanding the commodity frontiers that further demand exploitation and accumulation of resources by dispossessing local communities in order to preserve a high-demanding and consuming economy.

Elite international ENGOs, with a more focused approach on the influence of policy-making and the international negotiation process, have focused on addressing climate justice through predominant socio-economic configurations. They advocate for change within the prevailing relations of power under capitalism that determine the international climate negotiation process. Although there are some interactions between the international ENGOs and local grassroots movements, tensions persist in their relationship. These tensions are focused mainly over the discourse concerning capitalism and the use of market logics to address climate change. While ENGOs have supported the reform agenda that has been presented in the climate change negotiations, local grassroots movements have systematically moved away from that agenda (Bond 2011). This calls into question whether the international negotiation process can actually incorporate calls for justice while negotiating other issues, such as relative economic gains for national development and geopolitical objectives (Okereke and Coventry 2016).

Climate justice and the post-political

Despite the inability of the international negotiation process to yield actual climate justice, the Paris Agreement has been celebrated by some as a major victory in the fight against climate change (WRI 2016). In spite of the clear limited success of the negotiation process and the failures of the climate regime to reduce emissions globally (see IEA 2019), even after the announced withdrawal of the United States from the agreement in 2017, no optimism appears to have been lost in the international negotiations. Why has optimism in the process continued? One of the answers to this question lies in the fact that most of the decision-making processes in the climate change regime have been colonized by neoliberal, market logics that are being promoted as a means to solve climate change (Chatterton et al. 2013).

Dominant responses to address climate change have been driven by the recognition of a global carbon budget and the temperature limits set forth by the Intergovernmental Panel on Climate Change (IPCC) (IPCC 2018). The main solutions involve an acceleration of the global transition to renewable energy, a large-scale deployment of technological solutions, such as geo-engineering, with much emphasis placed on market solutions to address the problem (Chakrabarty 2017). Proponents of this logic, as exemplified by the Breakthrough Institute's 'Ecomodernist Manifesto', aim to accelerate technological progress as a means to achieve what they call a 'good Anthropocene' that will be capable of stabilizing the climate and protecting the natural world (Nordhaus et al. 2015). This, according to the manifesto, must be done by accelerating the development of technologies that could result in 'decoupling population and economic growth from environmental degradation' and by 'accelerating technological progress such as carbon capture and storage, nuclear power and desalination techniques' (Nordhaus et al. 2015).

Critics of this approach argue that these suggested solutions tend to render most of the discussion into purely technical and economic market-based solutions in which climate change is seen as a challenge for the development of technology, not as a social crisis. This has led to authors such as Erik Swyngedouw to claim that international climate governance has become a 'post-political populist project' where climate justice and other concerns over the environment are mobilized in such a way that the political dimension is suspended and suppressed (Swyngedouw 2011). The term 'post-political' refers to the progressive replacement of dissent and disagreement with widely disseminated techniques of management. This replacement rejects ideological divisions and reduces the political terrain to the sphere of consensual governance and policy-making centred on technical, managerial and consensual administration of environmental, social, economic or other domains (Swyngedouw 2011: 266). At the centre of this argument is the claim that a rapid substitution of fossil fuels by renewable energy, setting a correct price on carbon through taxes or markets, and decoupling carbon dioxide emissions from economic growth, will be enough to meet the Paris Agreement's targets (Caradonna et al. 2015).

These types of proposed solutions pose difficult questions from a distributive justice perspective. For example, how will the distribution of risks and benefits from these schemes be set between the rich and poor people, developed and developing countries, future and present generations, and humans and non-humans? Who will get to transition first, and how? How can we determine a 'correct' price on carbon? Who will pay? All of this ultimately poses the very necessary question: Are these the right solutions to climate change?

This post-political condition produces a consensual framework. It is a 'disavowal of dissensus and prevents antagonistic disagreement over real alternative socio-ecological futures by nurturing a widely populist and de-politicizing discourse and practice' (Swyngedouw 2017: 306). Consensual politics evacuate any real possibility to change or modify techno-managerial solutions to climate change. This is achieved by what Ernstson and Swyngedouw (2018), following on Roberto Esposito (2011), have called 'immunitary biopolitical governance' understood as 'a set of practices, rules, institutions and techno-managerial proceedings, that work to create an imaginary sense of protection and sequestration' (Ernstson and Swyngedouw 2018: 35). This fantasy of immunity serves to maintain the illusion that targeting carbon dioxide (CO_2) through a complex set of technological solutions, including geo-engineering, nuclear energy and other complicated schemes, such as emission trading and offsetting, can be used as 'simulative politics' giving the appearance of 'fixing the problem' without any substantial change to economic and political conditions (Blühdorn 2007).

Presenting climate justice as an empty signifier – as an idea that is 'subject to radically diverse interpretations and, as such, prevents anything concrete from being done' (Brown 2016: 116) – is one of the main results of the post-political condition. Climate justice is quickly devoid of meaning or content, thus allowing specific groups – those in power and with influential agendas – to temporarily fix its meaning and content, and to adopt particular solutions to the problem (Brown 2016). This perpetuates the post-political condition whereby international negotiations on climate change can almost exclusively deal with the removal of CO_2, ignoring the potential uneven distribution of costs and benefits of such actions and the already uneven power relations in which these actions are carried out.

In sum, the dominant climate justice movement has had an important weakness. The political is a 'contested social terrain where different imaginings of possible socio-ecological orders compete over the symbolic and material institutionalization of these visions' (Swyngedouw 2015: 90). Through the colonization of political space by international ENGOs and other elite actors, the international climate regime has incorporated justice demands (i.e., the recognition of different degrees of vulnerabilities associated with rights and responsibilities to mitigate GHG emissions), but it has systematically universalized and rendered all solutions as problems to be tackled through quick technological or market fixes, thus foreclosing any real debate over the heterogeneous demands of the wider climate justice movement.

Degrowth vs. 'green' growth

As the climate change debate was elevated into international negotiations, an important movement was emerging from the old argument about limits to growth that was popularized by the Club of Rome in the early 1970s (Meadows et al. 1972). The degrowth movement emerged in this context as an important field, constituted mainly of academics and social activists from developed nations (Martínez-Alier 2012). Indeed, degrowth is not a theory but an academic and social *movement*. It is first and foremost a socio-environmental movement that calls for a radical critique of the dogma of economic growth (Kallis 2018). The critique is that economic growth is undesirable and uneconomic because the cost of growth exceeds the benefits it produces (Kallis 2018). Economic growth is based on exploitation as it requires economic surplus that must be obtained by exploiting people (Kallis 2018). The quest for growth necessarily expands commodity frontiers, which in turn expands the scope of commodification of nature, social relations and products (Conde and Walter 2015).

Yet degrowth goes beyond a critique of economic growth. As a social movement, it puts forth a ruthless critique of capitalism and the constant need of the growth enterprise to commodify in order to produce surplus, which must then be reinvested to produce more surplus. This process, once set in motion, never stops of its own accord (Bellamy Foster et al. 2010). Degrowth implies a society beyond growth and a social direction where sharing, simplicity, conviviality, care and the commons are the basis for a voluntary transition towards a contraction-based economy in line with ecological limitations and greater social equity (Perey 2016). Hence, a paradoxical condition emerges: although there has been recognition of the impacts of climate change and the influence of capitalism in causing it, criticisms of capitalism tend to be silenced as proposals to address climate change are translated into a challenge for capitalism itself by advocating for (1) an absolute substitutability of everything in nature so that nothing natural is irreplaceable or irreversible; (2) a dematerialization or decoupling of the economy from resource use and GHG emissions; and (3) the conversion of nature into 'natural capital' whereby everything in nature is assigned an economic value (Bellamy Foster et al. 2010: 112).

These attempts to solve the contradictions of capitalism are more commonly known as 'green growth', which is under a broader set of practices that fit into the 'green economy' – the idea that the market can be steered towards environmental preservation if externalized costs of environmental degradation were internalized into prices (Sandberg et al. 2019). Most of these costs originate from the neoclassical view of economics, which argues that environmental degradation and climate change are to be understood as market failures. Solutions to this problem rest in technological fixes (deployed at a global scale), the correction of markets to incorporate and recognize their externalities (through policies such as carbon pricing) and maintaining the enterprise of economic growth and societal metabolism at its present rate (Sandberg et al. 2019). Under the discourse of green growth, technological innovation is capable of separating growth from environmental degradation, while at the same time keeping levels of consumption and productivity untouched (Jackson 2016).

However, green growth has proven ineffective at stopping environmental degradation. No evidence of absolute decoupling – that is, the situation whereby emissions decline in absolute terms, while economic output and growth continue – has actually taken place, while relative decoupling – the decline in emission intensity of economic output – has only occurred partially because of efficiency gains in the use of energy in some developed countries (Jackson 2016). At the same time, supporters of eco-efficiency solutions fail to recognize the role of the 'rebound effect' or 'Jevons Paradox', which show that 'efficiency gains normally lead to a decrease in the effective price of the commodity, thereby generating increasing demand, so that gains in efficiency do not produce a decrease in consumption to an equal extent' (Bellamy Foster et al. 2010: 177). This problem, which has been neglected by most economists and some environmentalists, demonstrates the fallacy of current notions that environmental problems can be solved merely by technological means.

Quick fixes: Promethean technologies and market optimisms?

Techno-optimism or Prometheanism is a belief in the idea that technological innovation is capable of bringing about the promise of transformation, which is necessary to surpass the limits imposed by the state of technological and social organization. The concept of sustainable development presented technology as something to be managed to maintain and promote an era of economic growth (Brundtland 1987: 16). Embedded into the post-political condition, Promethean technologies (or the promise of technology) are the ultimate resource and structure by which capitalism aims to surpass and expand despite ecological limits. Despite the lack of evidence of absolute decoupling (Kallis 2018), and the prevailing problem of the Jevons Paradox, one of the central proposals of the international climate negotiations process is rapid technological innovation and transfers that could maintain and support the purposes of a green economy and sustainable development. International NGOs, along with developing countries, have made explicit calls for technological transfers as a cornerstone of the Paris Agreement, and more so as an instrument to secure access to development and justice. Technology is seen as the main source for growth, decoupling and low-carbon development (Okereke and Coventry 2016, De Lucia 2012, Muraca 2012).

As part of the dominant narrative of eco-modernism, technological innovation and transfers are labelled as unproblematic and neutral solutions to climate change. They are assumed as being a simple substitution of sources that will allow a continuous use of resources. In other words, technological innovation is usually presented as a 'deus ex machina' (De Lucia 2012), a solution by which the problem of climate change can be solved without any meaningful change in the way that society and the international economy of capitalism is organized and structured. In other words, such approaches are seen collectively as a quick fix.

Technological transfers are normally placed as a necessary precondition to 'achieve sustainability' or as 'essential for low-carbon development'. Alf Hornborg's *The Power of the Machine* (2001) explicitly shows how technology rests on specific production and exchange relations that embody the past, present and future entanglements of power and reproduction of social and productive organizations of society, and explicitly shows how technologies are not value-neutral, but instead embody specific values and social relations into their value chains, services and products, making it extremely difficult to apply those technologies to uses that reflect different values (Huesemann and Huesemann 2011: 313).

In this context, it is impossible to separate technological transfers, such as renewable technologies, from particular embodiments of a specific economic, social and productive organization of society and nature (De Lucia 2012). For example, one only needs to look at the development of wind farms in Oaxaca, Mexico, where large-scale projects have enclosed common land and reproduced a specific political and discursive struggle over accumulation, all under the name of mitigating climate change (Boyer and Howe 2016, Avila-Calero 2017). In other words, technical solutions are normally presented as if completely removed from the world in which they operate, without any sense of the social and economic relations of power (Bellamy Foster et al. 2010: 84), thereby reproducing the capitalist project of transferring a mode of production and ideology into a mode of domination (De Lucia 2012). The associated injustices are perpetuated.

Technological innovation, along with carbon commodification, has emerged as a mainstream solution to address global climate change (Matt and Okereke 2011). Instruments such as cap-and-trade and offsetting mechanisms have been recognized as opportunities by important economic and finance industries (Böhm et al. 2012). Their adoption is again based on the misconception that capitalism can actually manage to sustain and adapt itself into a carbon-free logic (Newell and Paterson 2010). However, in reality carbon markets are part of the post-political condition, whereby CO_2 is presented as a new commodity producing new forms of dispossession by granting rights to emit it, and at the same time enclosing physical spaces and nature in the process of carbon offsetting (Böhm et al. 2012). This process can actually lead to new forms of environmental degradation in some spaces, while exacerbating enclosures of the commons in others.

Capitalism is prone to crises, as these are essential to the way it reproduces itself (Harvey 2014). It is through such crises that capitalism is able to expand commodity frontiers as a means to maintain the exploitation of economic surpluses (i.e., by exploiting labour, nature and land). Carbon markets can be seen as a further expansion of the commodity frontiers of capitalism in order to commodify CO_2 and perpetuate new forms of accumulation and the continuation of the post-political order in which offsetting allows for dubious forms of trading that would create 'the world's largest financial derivative market in the form of carbon trading' (Bellamy Foster et al. 2010: 428).

At the same time, presenting CO_2 as the only culprit of climate change is central in the strategy to legitimize the idea that capitalism and its political re-constitution of societies is unproblematic if CO_2 emissions are reduced. This process effectively displaces the association of climate change with the social metabolism imposed and expanded by capitalism and the uneven distribution (i.e., injustices) of power relations. Instead, it presents CO_2 as the problem that must be tackled through the existing norms, institutions and instruments that were set in place by capitalism itself (Swyngedouw 2017). This is the basis of how capitalism operates and perpetuates the post-political condition, whereas social organizations and present uneven (and unjust) distributions of power do not invite any change at all.

Therefore, green growth seems unlikely to provide the changes necessary to solve climate change. Criticisms to degrowth focus mainly on the fact that it is politically unfeasible, imprecise and socially undesirable. These criticisms are based on the fact that, so far, degrowth remains mostly isolated and lacks practical evidence (Sandberg et al. 2019). However, recent literature has shown that degrowth still presupposes a more coherent way to address environmental degradation and climate change (see: Kallis 2018, Kallis et al. 2018). Therefore, if CO_2 commodification and promises of technological optimism have proven to be ineffective fixes to address the climate conundrum, recognizing the biophysical limits presented by climate change calls for a radically different social metabolism (Sorman 2015).

The dominant discourse of technology is linked with economic growth, where innovation is normally presented as a precondition to sustain economic growth. New approaches from the degrowth movement and literature aim to (re)evaluate the role of technology as a potential tool for solidarity, sustainability and autonomy, instead of maintaining and reproducing hegemonic social relations under capitalism (see: Kerschner et al. 2017). The same can be said of the perpetual commodification of nature and CO_2 through carbon markets and cap-and-trade systems. These approaches tend to render the problem into technical solutions that apparently eliminate the need for political discussions and instead frame problems and their solutions as post-political.

Conclusion

The climate justice movement has been effective in making visible the injustices that have arisen from climate change, and it has highlighted the inequalities that appear from the actions taken to mitigate or adapt to this phenomenon. At the local level, struggles for climate (and environmental) justice around the world have been effective in giving a voice to displaced and disenfranchised communities affected by the progressive expansion of the commodity frontiers, and by the impacts of climate change (Schmelzer 2017, Muller 2015). The climate justice movement has also been effective in pointing out that, under the structure of capitalism, there are no universal, cheap and easily accessible solutions to the problem. In other words, addressing climate change will require more than treaties, financial transfers and technological innovation; it will also require democratic decision-making processes, local recognition and more effective distribution of the costs and benefits.

While much hope has been placed in the Paris Agreement, emissions have continued to rise steadily since its adoption, and under the current policy scenarios temperature limits will result in a 3.2 to 4°C temperature increase by 2100 (Mann and Wainwright 2018). The result of such warming will most likely produce an uneven world where some will be able to maintain a privileged lifestyle under a warmer condition, while others will have to either suffer the consequences or be forced to 'adapt' to capitalist exploitation and accumulation (Mann and Wainwright 2018). Consequently, the need for a radical climate justice movement is urgent. However, one of the main issues that have hindered the climate justice movement has been the lack of a coherent narrative and a unified position against capitalism and the expansion of commodity frontiers. While local movements are concerned with actions 'on the ground', the international negotiations and ENGOs have perpetuated the post-political condition and favoured approaches such as the green economy whereby techno-managerial and eco-modernism practices are framed as universal solutions to the problem of climate change. At the same time, while the academic approach towards climate justice has been more critical of the UNFCCC process, it also lacks a coherent narrative to place climate justice directly opposed to capitalism.

Degrowth offers three main lessons for the climate justice movement. First, it demands a radical criticism of the dogma of growth, the reformist ideas that aim for 'green' capitalism and the social imaginaries that call for growth as a necessity in modern societies. Degrowth shows that calls for climate justice must necessarily criticize the depoliticizing narratives of capitalism and its approach towards universal solutions to climate change. Climate justice is then a *political* issue that cannot be solved by focusing on technical fixes and managerial solutions.

Second, degrowth produces a common language through which opposition to commodification, exploitation and accumulation are present in local and grassroots movements. Under the banner of degrowth, understood as a political slogan, movements across the world are starting to resist and oppose the progressive expansion of capitalism's traits, such as accumulation by dispossession, as exemplified (for example) by the expansion of extractive projects such as fracking, open-pit mining and land occupations for 'renewable' energy projects (Mann and Wainwright 2018). These social movements are now articulating a common ground and awareness of the problem, despite most of them remaining unarticulated and politically unrepresented in the international negotiations on climate change.

Finally, degrowth injects a new sense of urgency and hope. Concepts like 'nowtopias' (Carlsson 2015) argue for redefinition of the values of society based on that which is undervalued by the market, and they aim for the reorganization of society and collective action beyond the market and the state. Degrowth offers a way forward towards a society based on solidarity, autonomy and a slower metabolism. As a social and academic movement, degrowth has served as a platform to articulate alternatives of the future that go beyond growth, development and capitalism. It has opened up spaces for imagining new ways of achieving convivial societies that can lead to prosperity (Kallis 2018).

Overall, there is an urgent need for the climate justice movement to engage with the political and to challenge the logic of growth and capitalism; to pay closer attention to how local communities and movements are articulated against the expansion of capitalism's commodity frontiers; and to incorporate degrowth's critique and alternatives into the social metabolism of capitalism in order to move away from techno-optimism and carbon commodification towards a more democratic and equitable society. Otherwise, radical demands for climate justice will continue to be colonized by a perpetual post-political condition that will forever loom over the climate justice movement, leaving the realization of climate justice beyond reach.

Acknowledgements

My thanks go to Paul G. Harris for all his support and patience. Thanks to the Mexican Climate Initiative (ICM) and the Department of Sustainability Studies at Iberoamericana University, Mexico City, for their support and for granting me the time to write this chapter. Thanks to Ana Sofia Tamborrel and Jorge Villarreal and to Ximena Estavillo for her love and support in writing this chapter.

References

Avila-Calero, Sofia (2017), 'Contesting energy transitions in Mexico: wind power and conflicts in the Isthmus of Tehuantepec', *Journal of Political Ecology*, **24**, 992–1012.

Bellamy Foster, John, Brett Clark and Richard York (2010), *The Ecological Rift: Capitalism's War on the Earth*. New York: Monthly Review Press.

Blühdorn, Ingolfur (2007), 'Sustaining the unsustainable: symbolic politics and the politics of simulation', *Environmental Politics*, **16**(2), 251–75.

Böhm, Stephan, Maria Ceci Misoczky and Sandra Moog (2012), 'Greening capitalism? A Marxist critique of carbon markets', *Organization Studies*, **33**(11), 1617–38.

Bond, Patrick (2011), 'Politics of climate justice. Paralysis above, movement below', paper presented at the Gyeongsang University Institute of Social Science, Jinju, 27 May.

Boyer, Cymene and Dominic Howe (2016), 'Aeolian extractivism and community wind in southern Mexico', *Public Culture*, **28**(2(79)), 215–35.

Brown, Trent (2016), 'Sustainability as empty signifier: its rise, fall, and radical potential', *Antipode*, **48**(1), 115–33.

Brundtland, Gro Harlem (1987), *Our Common Future*, The World Commission on Environment and Development, Oxford: Oxford University Press.

Bulkeley, Harriet, Gareth A.S. Edwards and Sara Fuller (2014), 'Converting climate justice in the city: examining politics and practice in urban climate change experiments', *Global Environmental Change*, **25**, 31–40.

Caradonna, Jeremy, Iris Borowy, Tom Green, Peter A. Victor, Maurie Cohen, Andrew Gow, Anna Ignatyeva et al. (2015), 'A call to look past an ecomodernist manifesto: a degrowth critique', *Resilience.org*. Accessed 12 January 2019 at: https://www.resilience.org/wp-content/uploads/articles/General/2015/05_May/A-Degrowth-Response-to-An-Ecomodernist-Manifesto.pdf.

Carlsson, Chris (2015), 'Nowtopians', in Giacomo D'Alisia, Federico Demaria and Giorgos Kallis (eds), *Degrowth: A Vocabulary for a New Era*, New York, NY, USA and London, UK: Routledge Publishing, pp. 215–17.

Chakrabarty, Dipesh (2017), 'The politics of climate change is more than the politics of capitalism',

Special Issue: Geosocial Formations and the Anthropocene, *Theory, Culture and Society*, **34**(2–3), 25–37.

Chatterton, Paul, David Featherstone and Paul Routledge (2013), 'Articulating climate justice in Copenhagen: antagonism, the commons, and solidarity', *Antipode*, **45**(3), 602–20.

Conde, Marta and Mariana Walter (2015), 'Commodity frontiers', in Giacomo D'Alisia, Federico Demaria and Giorgos Kallis (eds), *Degrowth: A Vocabulary for a New Era*, New York, NY, USA and London, UK: Routledge Publishing, Taylor and Francis Group, pp. 71–4.

De Lucia, Vito (2012), 'The climate justice movement and the hegemonic discourse of technology', in Matthias Dietz and Heiko Garrelts (eds), *Routledge Handbook of the Climate Change Movement*, New York, NY, USA and London, UK: Routledge, pp. 66–83.

Ernstson, Henrik and Erik Swyngedouw (2018), 'O tempora! O mores! Interrupting the anthropo-obScene', in Henrik Ernstson and Erik Swyngedouw (eds), *Urban Political Ecology in the Anthropo-Obscene: Interruptions and Possibilities*, New York, NY, USA and London, UK: Routledge City Series, pp. 25–47.

Esposito, Roberto (2011), *Inmmunitas*, Cambridge: Polity Press.

Harris, Paul G. (2010), 'Misplaced ethics of climate change: political vs. environmental geography', *Ethics, Place and Environment*, **13**(2), 215–22.

Harris, Paul G. (2016), *Global Ethics and Climate Change*, Edinburgh: Edinburgh University Press.

Harvey, David (2014), *Seventeen Contradictions and the End of Capitalism*, London: Profile Books.

Hornborg, Alf (2001), *The Power of the Machine: Global Inequalities of Economy, Technology, and Environment*, Walnut Creek, CA: AltaMira Press.

Huesemann, Michael and Joyce Huesemann (2011), *Techno-Fix: Why Technology Won't Save Us or the Environment*, Gabriola Island: New Society Publishers.

Intergovernmental Panel on Climate Change (IPCC) (2018), 'Summary for policymakers of IPCC's special report on global warming of 1.5°C'. Accessed 10 January 2019 at: https://www.ipcc.ch/site/assets/up loads/2018/11/pr_181008_P48_spm_en.pdf.

International Energy Agency (2019), *Global Energy and CO2 Status Report. The Latest Trends in Energy Emissions in 2018*. Accessed 26 March 2019 at: https://webstore.iea.org/global-energy-co2-status -report-2018.

Jackson, Tim (2016), *Prosperity Without Growth: Foundations for the Economy of Tomorrow*, New York, NY, USA and London, UK: Routledge Publishing.

Kallis, Giorgos (2018), *Degrowth*, New York: Agenda Publishing.

Kallis, Giorgos, Vasilis Kostakis, Steffen Lange, Barbara Muraca, Susan Paulson and Matthias Schmelzer (2018), 'Research on degrowth', *Annual Review of Environment and Resources*, **43**, 291–316.

Kerschner, Christian, Petra Wächter, Linda Nierling and Melf-Hinrich Ehlers (2017), 'Degrowth and technology: towards feasible, viable, appropriate and convivial imaginaries', *Journal of Cleaner Production*, **197**(2), 1619–36.

Mann, Geoff and Joel Wainwright (2018), *Climate Leviathan: A Political Theory of Our Planetary Future*, London, UK and New York, NY, USA: Verso Books.

Martínez-Alier, Joan (2012), 'Environmental justice and economic degrowth: an alliance between two movements', *Capitalism Nature Socialism*, **23**(1), 51–73.

Martínez-Alier, Joan, Leah Temper, Daniela Del Bene and Arnim Scheidel (2016), 'Is there a global environmental justice movement?', *The Journal of Peasant Studies*, **43**(3), 731–55.

Matt, Elah and Chukwumerije Okereke (2011), 'A neo-Gramscian account of carbon markets: the case of the European Union Emissions Trading Scheme and the Clean Development Mechanism', in Benjamin Stephan and Richard Lane (eds), *The Politics of Carbon Markets*, London: Routledge, pp. 113–32.

Meadows, Donella H., Dennis L. Meadows, Jorgen Rander and William Behrens (1972), *The Limits to Growth*, New York: Universe Books.

Muller, Tadzio (2015), 'Climate justice and degrowth: a tale of two movements'. Accessed 25 January 2019 at: https://www.degrowth.info/en/2015/03/climate-justice-and-degrowth-a-tale-of-two-movements/.

Muraca, Barbara (2012), 'Towards a fair degrowth-society: justice and the right to a "good life" beyond growth', *Futures*, 44, 515–45.

Newell, Peter and Matthew Paterson (2010), *Climate Capitalism: Global Warming and the Transformation of the Global Economy*, Cambridge: Cambridge University Press.

Nordhaus, Ted, John Asafu-Adjaye, Linus Blomqvist, Stewart Brand, Barry Brook, Ruth Defries, Erle Ellis et al. (2015), 'An ecomodernist manifesto'. Accessed 25 November 2018 at: http://www.ecomodernism. org.

Okereke, Chukwumerije and Philip Coventry (2016), 'Climate justice and the international regime: before, during and after Paris', *Wiley Interdisciplinary Reviews: Climate Change*, 7(6), 834–51.

Perey, Robert (2016), 'Degrowth as a transition strategy', in H.G. Washington and P. Twomey (eds), *A Future Beyond Growth: Towards a Steady State Economy*, Abingdon: Routledge, pp. 213–22.

Sandberg, Maria, Kristian Klockars and Kristoffer Wilén (2019), 'Green growth or degrowth? Assessing the normative justification for environmental sustainability and economic growth through critical social theory', *Journal of Cleaner Production*, 206, 133–41.

Schlosberg, David and Lisette B. Collins (2014), 'From environmental to climate justice: climate change and the discourse of environmental justice', *Wiley Interdisciplinary Reviews: Climate Change*, 5(3), 359–74.

Schmelzer, Matthias (2017), 'No degrowth without climate justice', *Degrowth Info*. Accessed 20 November 2018 at: https://www.degrowth.info/en/2017/02/no-degrowth-without-climate-justice/.

Sorman, Alevgül (2015), 'Societal metabolism', in Giacomo D'Alisia, Federico Demaria and Giorgos Kallis (eds), *Degrowth: A Vocabulary for a New Era*, New York, NY, USA and London, UK: Routledge Publishing, pp. 41–4.

Swyngedouw, Erik (2011), 'Depoliticized environment: the end of nature, climate change and the post-political', *Royal Institute of Philosophy Supplement*, 69, 253–74.

Swyngedouw, Erik (2015), 'Depoliticization (the political)', in Giacomo D'Alisia, Federico Demaria and Giorgos Kallis (eds), *Degrowth: A Vocabulary for a New Era*, New York, NY, USA and London, UK: Routledge Publishing, pp. 90–93.

Swyngedouw, Erik (2017), 'CO2 as neo-liberal fetish: the love of crisis and depoliticized immuno-biopolitics of climate change', in Damien Cahill, Melinda Cooper, Martijn Konings and David Primrose (eds), *The Sage Handbook of Neoliberalism*, London: Sage Publishing, pp. 295–314.

WRI (2016), 'Statement: Paris agreement crosses final threshold to enter into force', blog entry by Lauren Zelin and Rhys Gerholdt. Accessed 15 January 2019 at: https://www.wri.org/news/2016/10/statement-paris-agreement-crosses-final-threshold-enter-force.

WWF (2016), 'Paris agreement. What it is, why it matters, and where we go from here in the fight against climate change'. Accessed 20 January 2019 at: https://www.worldwildlife.org/pages/paris-climate-agreement.

7 A cosmopolitan agenda for climate justice: embracing non-state actors

Alix Dietzel and Paul G. Harris

A cosmopolitan viewpoint is essential for fully understanding the complexities of the climate change problem and finding effective and just solutions to it (Harris 2010, 2011a, 2016). Nation-states, which have dealt with most global problems in the past – war and peace, national and international development, global health and human rights – have not been up to the task of dealing with climate change effectively or justly. With global greenhouse gas emissions *continuing to increase* (Pierre-Louis 2018), global temperatures rising and climatic changes taking hold around the world, it is increasingly apparent that not enough is being done. Although states have started to respond to climate change, for example by gradually implementing their pledges under the 2015 Paris Agreement, their actions have so far been agonizingly slow-paced and grossly inadequate. Indeed, in May 2019 the United Nations Secretary General declared that 'climate change is running faster than our efforts to address it – and political will in many parts of the world is unfortunately slowing down' (Guterres 2019). In response, scholars, policy makers and activists alike are increasingly focusing their attention on an 'alternative response' among non-state actors (sometimes referred to as sub-state, networked or transnational actors).

In contrast to dominant approaches to climate politics and justice – which tend to focus on the roles of nation-states and relations among them – cosmopolitanism by definition is theoretically and morally blind to national borders. As such, it can open our eyes to the roles played by individuals and other non-state actors, regardless of their nationality. A cosmopolitan agenda for climate justice would move beyond considering what states can and should do about climate change to consider instead what role non-state actors ought to play in a just response to it. Non-state actors, including cities, corporations, non-governmental organizations and individuals, have enormous potential to contribute to the climate change solutions (Hsu et al. 2015). Cosmopolitanism can highlight that potential.

The role of non-state actors is especially interesting at this time because the Paris Agreement presented a significant shift away from a purely state-led approach in global climate governance. The post-Paris climate regime does not see non-state actors as merely a 'helpful addition' to the actions of states, but rather as a core element of action (Hale 2016: 14). It is important to take stock of this development and to reflect on what it means for the worldwide response to climate change. Will the climate change response become more efficient, or more chaotic, with greater involvement of non-state actors? Will non-state actors inspire states to do more, or will states become reliant on non-state responses? The emergence of non-state action also raises important ethical questions, in particular about whether non-state actors could play a role in realizing a just response to climate change (Dietzel 2019: 2). Here, we make the case for a cosmopolitan climate justice agenda that explores these kinds of questions. We argue that non-state actors, including individuals, have significant potential to contribute to a just response to climate change. Focusing on and assessing these actors should form a key part of the climate justice agenda.

In this chapter we start by describing the cosmopolitan approach to climate justice and argue that it is needed for understanding the complexities of the climate change problem. Cosmopolitan justice and climate justice go hand in hand; you cannot have the latter without the former. Next, we discuss why moving beyond states is so important, exploring the changing policy context and the potential of non-state response, before delineating the roles that individual and collective non-state actors can play in a just response to climate change. We argue that non-state actors have the potential to contribute to global mitigation and adaptation efforts, and to significantly shift the context of the climate change response. We encourage a research agenda for climate justice that takes a cosmopolitan viewpoint.

Cosmopolitan climate justice

Cosmopolitanism is a rich and varied discipline, but most cosmopolitan scholars take the individual as their moral starting point. They assume that all human beings have equal moral worth, and therefore that they have the right to equal moral consideration (see Harris 2011b, Harris 2016: 103–107). From a cosmopolitan viewpoint, individuals have moral worth regardless of nationality. Indeed, for die-hard cosmopolitans, national borders have no moral significance at all. (Naturally, cosmopolitans are not blind to the tremendous practical significance of those borders.) As David Held has argued, cosmopolitans want to draw our attention to 'the ethical, cultural, and legal basis of political order in a world where political communities and states matter, but not only and exclusively', noting that in a world where states are 'tightly intertwined, the partiality, one-sidedness and limitedness of "reasons of state" need to be recognized' (Held 2005: 10). Many cosmopolitans go further, arguing more broadly for a 'refusal to regard existing political structures as the source of ultimate value' (Barry 1999: 36). For cosmopolitans, there are global responsibilities and values that exist irrespective of state borders, and in principle everyone has a duty to care about things that are happening in other places (Dower 2007: 11, 28).

The focus on the moral importance of the individual has led some cosmopolitans to critically engage with theories of justice that go behind traditional preoccupation with the state. Such endeavours have led to the emergence of the discipline of cosmopolitan or 'global' justice in which scholars argue that questions of justice ought to be global in scope (Harris 2016: 101–22). Scholars of cosmopolitan justice focus on what individuals across the world deserve and how distribution of deserved entitlements can be achieved, as well as considering the role individuals can play and challenging attempts to relieve them of their moral responsibility. Many of these scholars concern themselves with problems of global cohabitation in which individuals are not yet treated as morally equal, or where the moral focus has been exclusively directed towards states. The focuses of such scholarship include global poverty (Brock 2009), humanitarian intervention (Archibugi 2004), health (Ooms and Hammonds 2010), gender inequality (Pepper 2014) and refugees (Gibney 2015). In keeping with the fundamentals of cosmopolitanism, scholars of cosmopolitan justice view the national community (i.e., the state) as the incorrect moral basis for determining that which is just. For them, justice ought to obtain globally (not just nationally).

As responses to climate change have increasingly lagged behind what scientists recommend, some cosmopolitan justice scholars have become interested in climate change and its (mis)management (Caney 2014, Dietzel 2017, 2019, Harris 2008, 2010, 2016, Heyward and Roser 2016, Maltais and McKinnon 2015, Lawrence 2014, Vanderheiden 2008). Cosmopolitans often base their argument for global justice on either (or both) of two arguments (the first exemplified by Beitz 1999): that levels of international cooperation are now so great as to constitute a global community in which conceptions of justice that traditionally applied only within communities now apply globally; and that globalization (economic, social, environmental and so forth) and interdependence are now so great that people everywhere affect people everywhere else, often in dramatic ways (Harris 2016: 107). This is particularly true with respect to climate change because it is manifested by interdependence across national borders, and indeed irrespective of them, and it is caused by pollution arising from every populated place on the planet, with its impacts being felt everywhere, often in the extreme.

Cosmopolitan thinking on justice is ideally placed for exploring the complexities of the climate change problem. This is because climate change is a problem of cosmopolitan justice by its very nature. Greenhouse gas emissions pollute the global atmosphere and have impacts well beyond the borders of individual states. In this sense, 'the global nature of the climate change problem defies conventional assumptions about state sovereignty and geographically bounded justice' (Vanderheiden 2008: xiv). In addition, individuals are the ultimate causes and victims of climate change: any individual on the planet can cause emissions that will harm any individual in any other part of the world (Dietzel 2019: 14). This raises distributive questions about which individuals should be protected and which individuals may have to refrain from emitting. The cosmopolitan climate justice approach is primed for addressing these questions head on because it focuses on the individual as a morally important agent. This allows for a normative exploration of the role and responsibilities of individuals (and groups of them) that cause climate change, based on the understanding that it is morally important to protect and compensate individuals suffering from climate change wherever they happen to live (Dietzel 2019: 15). In other words, cosmopolitan justice captures both the normative and practical realities of who causes climate change and who is affected by it (Harris 2016: 185–95).

Furthermore, cosmopolitan climate justice scholars are well placed to explore the extreme levels of unfairness among individuals across the globe that result from climate change. Climate change is imposed on people who are already poor, cannot adequately protect themselves and have little or no real say in the matter. It will most negatively affect the world's poor, including the majority of people living in the Global South. They are least able to cope with the impacts of climate change and they have done the least to contribute to the causes of it. In contrast, most of the people living in the Global North have contributed more to the problem, and they are more capable of coping with the impacts. Consequently, most of them will suffer the least (Dietzel 2019: 7). This is why the empirical conditions of climate change 'cry out for justice' (Harris 2010: 37), and this is why climate change quite naturally fits into the remit of cosmopolitan justice. Because cosmopolitanism is concerned with questions of fair distribution among individuals across the globe, it is well placed to explore who must do what and why given the severe inequalities between morally equal individuals.

Combating climate change will require a collaborative effort among a variety of actors; states have demonstrated that they cannot do it alone. More effective action on climate change will inevitably involve questions about which actors must lower emissions by how much, and which actors should contribute to the costs of addressing climate change. Cosmopolitan climate justice scholars can (and do) tackle such questions head on. Engaging in cosmopolitan climate justice debates means exploring ethical aspects of the climate change problem – identifying victims of injustice, defining a fair distribution of climate responsibilities, and assigning duties of justice to those who are responsible (Dietzel 2019: 1). These are all key aspects of the climate change problem that cannot be ignored if we are to develop a fair and effective global response. In addition, by its very nature, cosmopolitanism reminds us that non-state actors must play their part in realizing climate justice. A cosmopolitan approach to climate justice is therefore useful for understanding the complexities of the climate change problem and for putting forward policy recommendations that are focused on the realities of burden sharing among people around the world (Harris 2011a).

Moving beyond states in responses to climate change

In the search for climate justice, scholars and practitioners have often focused on multilateral (i.e., state-to-state) rather than non-state responses to climate change. For example, the principle of common but differentiated responsibility, which has served as a nominally accepted foundation of climate justice for more than a quarter of a century, is premised on the responsibilities of states (Harris 2016: 81–9). This is problematic because non-state actors, including cities, corporations, non-governmental organizations and individuals, have become increasingly important in responding to climate change, not least because states have done far too little on their own. Non-state actors can mitigate climate change, help affected communities adapt to climate impacts and leverage financial and other resources to fund both mitigation and adaptation measures (Sander et al. 2015: 467). In recent years there has been a 'Cambrian explosion' of non-state responses to climate change (Abbott 2012: 571); it is estimated that there are now several thousand in existence (Stevenson and Dryzek 2014: 87). The non-state response presents an opportunity to address climate change effectively (Bulkeley et al. 2015) and justly (Dietzel 2019).

The international climate regime itself has begun to take notice of the importance of non-state actors: they were formally invited to be part of multilateral negotiations for the first time at the 2013 nineteenth conference of the parties (COP19) to the Framework Convention on Climate Change in Lima, Peru. Those negotiations led to, among other outcomes, the creation of the Non-State Actor Zone for Climate Action, a platform that officially recognizes and tracks non-state climate change initiatives. To some observers, this indicated that the role of non-state actors was finally being taken seriously by multilateral actors (Hale 2016). At the 2015 COP21 negotiations in Paris, the conference president declared that non-state action was the 'fourth pillar' of the climate regime, 'alongside, and equal to, the conference's national pledges, its financing package and the negotiated agreement' (Hale 2016: 14). This was visible in the Paris Agreement itself, which noted 'the importance of the engagements of all levels of government and various actors' (UNFCCC 2015: 21). The recognition of non-state actors presents a significant shift in attitude among multilateral actors, demonstrating their importance in the global climate change regime generally and as part of the post-Paris Agreement response going forward.

It is vital to recognize the significance of these recent changes. It is now clear that failing to explore and assess non-state responses alongside international responses by states would ignore the realities of the global response to climate change. That said, while non-state actors are no doubt playing an increasingly significant role, the international climate regime is by no means losing momentum. If anything, it is growing in importance as many states continue to commit to significant mitigation and adaptation targets under the Paris Agreement. Assessing the international responses of states to climate change therefore remains important for cosmopolitan climate justice scholars, even as they explore the critical role of non-state actors in enabling a more just response (Diezel 2019: 3).

The vital role of individuals

A cosmopolitan approach to climate justice forces us to stop focusing exclusively on the role of states, thereby revealing the wide range of non-state actors that contribute to the problem and have the potential to mitigate it. Cosmopolitanism goes further, by its very nature, in drawing attention to those actors that ultimately cause climate change and are affected by it: individuals. Cosmopolitans argue that individual persons are moral agents with duties to behave in certain ways if they have the ability to do so. It follows from this that nearly all of the world's affluent people have such duties, because their affluence gives them capability and thus agency. The collective contribution of affluent individuals is staggering: the top 10 per cent of wealthiest individuals account for 49 per cent of global 'lifestyle consumption' emissions (Oxfam 2017). The impact of these actors changing their behaviour would be significant, and their capability to do so has been well established (Dietzel 2019: 84–5). In this sense, it is clear from a cosmopolitan perspective that these individuals have a moral duty to act. Furthermore, if there is a duty to address climate change and its injustices, then all affluent (capable) people share this duty.

The importance of this argument is potentially profound because, if it is accepted, it means that billions of people ought to be acting to address climate change. Very importantly, this requirement would apply to affluent people everywhere, including most people in the Global North but also hundreds of millions of affluent people living in the Global South (Harris 2016: 188–93). The fact that affluent people have capability to aid those who are suffering from climate change is arguably enough to justify the requirement that they behave responsibly. But it is also the case that most affluent people, including millions of them in the developing world, have benefited from past (historical) greenhouse gas pollution. For example, while most people of the Global North benefit from past pollution, it is also true that millions of affluent people in countries of the Global South do so as well through the wealth that comes from exports purchased by the people of the Global North (who are able to do so partly due to their own or their ancestors' historical pollution). All of these obligations obtain regardless of whether governments live up to their obligations under the Paris Agreement or some other measure. There is also an argument to be made about capable actors in an unjust system. The moral basis for this obligation stems from the wider system of injustice that is being created through cumulative emissions, which threatens individuals everywhere. Cosmopolitans argue that affluent individuals have a moral responsibility not to perpetuate or contribute to such an unjust system. Instead, they ought to change their behaviours within it or change the system itself if they are capable of doing so (Dietzel 2019: 83). This moral responsibility applies no matter where an affluent individual happens to reside.

From a cosmopolitan perspective, then, it becomes apparent that billions of individuals ought to be behaving in ways that are relatively rare at present. Of course, individual consumers are only able to choose from so many options for essential products and services. Some are simply not available, for example clean, efficient, affordable and safe public transport (Seyfang 2005: 297). In addition, individuals are often left out of decisions that make a major impact on emissions because those decisions are made on a societal, not individual level (Seyfang 2005: 296).

However, even if consumers are locked into certain choices, it would be impossible to argue that individual consumers have no way of altering their behaviour to affect emissions. Take one aspect of consumption as an example: food. About 25 per cent of global greenhouse gas emissions come from food production (Vermeulen et al. 2012). A study of British diets (Scarborough et al. 2014) found that the average daily greenhouse gas emissions in kilograms were 7.19 for high meat-eaters, 4.67 for low meat-eaters, 3.91 for fish-eaters, 3.81 for vegetarians and 2.89 for vegans. In other words, emissions were twice as high for meat eaters as for vegans. This means that those who already eat high levels of meat (over 100 grams per day) are capable of reducing their emissions significantly by reducing their consumption. To be sure, changing diet is most realistic when one has the means to do so – the capability to spend money on a nutritious vegan diet, for example. This differentiation in capabilities should not be ignored and must be considered when deciding which individuals should be held responsible for climate change action (Dietzel 2019: 85).

Individuals should start by doing what they can to reduce their direct and indirect emissions of greenhouse gases. This would require lifestyle changes (which could mean improving health and happiness) by, for example, riding bicycles instead of driving cars, or working from home instead of commuting to work. The quickest way for most affluent individuals to lower their greenhouse gas emissions is probably to stop (or greatly reduce) eating most foods derived from animals (especially those that are unhealthy, such as processed red meats), to stop flying unless it is unavoidably necessary and, when available, to use public transport exclusively. When possible (notably in democracies), individuals also ought to do what they can to influence political and policy processes that foster the elimination of fossil fuels from the energy mix. Furthermore, affluent individuals ought to be doing what they can, and encouraging their governments to do what they can, to aid those people around the world who are or will be negatively impacted by climate change, even if those people live outside one's own national community (i.e., state). Until governments do enough of this, affluent individuals can do so most easily by giving money to non-governmental organizations with programmes for climate adaptation (e.g. Oxfam).

Having said this, even cosmopolitans would acknowledge that actions by states – infrastructure, regulations, taxes, education programmes and the like – are important for fostering individual action. Indeed, actions by other non-state actors, such as corporations and non-governmental organizations, can (and need to) also facilitate and greatly supplement actions by individuals. After all, not all individual actors will (or can) act as they should. Realizing climate justice requires action by individuals, but it also requires it from other non-state actors.

Climate justice and collective non-state actors

A just response to climate change requires, at minimum, a substantial lowering of global emissions and a well-funded adaptation strategy to ensure that communities and individuals around the world are protected from climate change effects (Dietzel 2019: 66). Non-state actors – especially cities, regions, corporations and non-governmental organizations – not only have the potential to contribute to global mitigation and adaptation efforts, but they also have the potential to shift the context of the climate change response significantly and thereby make possible the systemic changes that are required to realize climate justice. Such a systemic shift will be required if individuals are to live up to the climate responsibilities outlined above. A cosmopolitan viewpoint can illuminate these important capabilities of non-state actors.

Lowering global emissions

Most non-state initiatives are not aimed at reducing greenhouse gas emissions directly (Dietzel 2019: 166). Instead, non-state actors usually focus on providing information, incentives and capacity building for other actors to reduce their emissions. Emissions reductions from these kinds of activities are difficult to measure. For example, a non-state project might focus on energy efficiency, changes to transportation and infrastructure, or development and deployment of green technology. All of these efforts will have an indirect impact on emissions levels, but it is difficult to pinpoint which action influenced how much of a reduction in emissions. Furthermore, even when non-state actors explicitly aim to reduce emissions, they rarely set specific emissions targets (Hsu et al. 2015). In the rare cases that non-state actors do set out specific targets, it is difficult to measure whether these targets have been met. Researchers must often rely on self-reporting because non-state actors are rarely monitored by a third party. For this reason, although there is some evidence of leadership on climate action, evidence of concrete measurable impacts on emissions is inadequate to reach definitive conclusions (Bulkeley et al. 2015: 161).

That said, recent studies on the subject are optimistic. For example, one study found that 238 leading city initiatives around the world could reduce emissions equal to those of all OECD countries in 2010 (Sander et al. 2015: 467). Similarly, a variety of small-scale studies released in the lead-up to COP21, looking only at a handful of the initiatives that bring together major cities, large corporations, renewable energy projects and other such actors, 'found their mitigation potential to be in the range of 2.5–4 billion tons of CO_2 by 2020, more than India emits in a year' (Hale 2016: 13). Corporations in particular are capable of lowering emissions substantially. The private sector accounts for more than one-third of energy consumed worldwide, and many corporations emit significantly more greenhouse gases than do major cities (Bulkeley and Newell 2010: 87). By lowering their own emissions, corporate actors could therefore make a significant impact on global mitigation efforts. Corporations also have an important role to play beyond targeting their own emissions. For example, groups of corporations have been found to effectively institutionalize new corporate norms, for example to disclose their carbon emissions (Pattberg 2010: 159). In addition, corporations can finance mitigation measures; it is estimated that 98 per cent of global investment and financial flows required to tackle climate change will need to come from the private sector (Bulkeley and Newell 2010: 88). Furthermore, corporations develop and disseminate most of the world's technology (Bulkeley and Newell 2010: 88). Although corporations have a very large role in causing climate change, they also have the potential to be a large part of the solution.

The Climate Group, made up of over one hundred major corporations, sub-national governments and international institutions, is an example of the role that corporations, along with other non-state actors, can play in limiting global emissions. The Climate Group has called for 'a clean revolution through the rapid scale-up of low carbon energy solutions that can be replicated worldwide' (Climate Group 2019b). Members of the Climate Group represent a significant amount of global wealth: the combined revenue of its corporate members is in excess of US$1 trillion (Climate Group 2019a). In this sense, there is great potential to make a difference by financing mitigation measures and implementing these on a large scale. To get a sense of this potential, consider one of their initiatives: the LED Lighting Project. In place in major cities, including London, New York, Hong Kong and Mumbai, this project has demonstrated that switching to LED street lights in cities can present energy savings as high as 80 per cent (Climate Group 2012: 24). If the project were implemented worldwide, it could potentially lower global carbon emissions by almost 5 per cent (because street lighting accounts for 6 per cent of global emissions), the equivalent of emissions from 70 per cent of the world's passenger vehicles. Through projects such as this, the Climate Group says that it is an 'advocate for stringent global action by demonstrating the availability of solutions' (Hoffman 2011: 85). It highlights the potential of non-state actors, notably corporations, to contribute to global mitigation efforts.

Financing adaptation to climate change

Non-state actors appear to strongly favour mitigation over adaptation measures. Bulkeley et al. found that 75 per cent of non-state projects focus on mitigation alone, while 22 per cent focus on both mitigation and adaptation, and 3 per cent focus solely on adaptation (Bulkeley et al. 2015: 130). However, Bulkeley et al.'s study focused on initiatives during the period of late 2008 to early 2010. Attitudes among non-state actors towards adaptation may have shifted since then, especially when considering that adaptation only recently emerged as a key concern among them (Hoffman 2011: 40). Adaptation could become an increasingly important component of the non-state response to climate change in the future (Dietzel 2019: 181). In fact, more recent studies suggest that non-state actors have the potential to help affected communities adapt to climate change, build capacity for adaptation, catalyse supportive political coalitions and deliver policy innovation (Sander et al. 2015: 467).

To explore the potential for adaptation in more detail, consider the role of another type of non-state actor: cities. It is estimated that two-thirds of the world's population will live in cities by 2030 (Hoffman 2011: 104). It is therefore crucial that they be involved in the process of adapting to climate change. Encouragingly, cities are already contributing to global adaptation efforts. By implementing policies at political scales significantly smaller than those of states, cities are able to demonstrate community-specific methods for coping with climate change (McKendry 2016: 1357). Cities can also demonstrate how local adaptation measures can be financed and implemented. Furthermore, localized climate adaptation efforts allow those affected to have a more direct say on how they are protected (McKendry 2016: 1357). This is important because realization of participatory justice is difficult to achieve in interstate climate change negotiations where local interests are rarely represented.

An example of the role that cities can play in global adaptation efforts can be found in the Asian Cities Climate Change Resilience Network (ACCCRN), which aims to 'strengthen the capacity' of more than 50 rapidly urbanizing cities in Bangladesh, India, Indonesia, the Philippines, Thailand and Vietnam so that they may 'survive, adapt, and transform in the face of climate-related stress and shocks' (ACCCRN 2019a). The ACCCRN covers more than 2.5 billion people and has dozens of city-based projects in place, suggesting significant scope for contributing to global adaptation efforts. Importantly for those concerned about achieving climate justice, the ACCCRN hopes to 'give marginalised communities an increasingly prominent voice in their cities' future through localised adaptation projects' (ACCCRN 2019c). One such project is the long-term flood resilience initiative currently taking place in Hat Yai, Thailand. This project involves planning and implementation of integrated flood risk-reduction measures (ACCCRN 2019b). The plan is for these measures to be developed and endorsed by local actors (private citizens and industry experts alike) on the ground before they are scaled up (ACCCRN 2019b). Projects such as this illustrate why non-state initiatives have the potential to promote localized adaptation measures and give voice to those most affected by climate change.

Restructuring the response to climate change

Directly lowering emissions and contributing to adaptation efforts are essential to just responses to climate change. But non-state actors might have an even greater indirect impact on the *context* of climate policies by 'transforming prevailing patterns of social and economic practice, generating novel forms of investment, city planning, business practice or even changing personal behaviour' (Bulkeley et al. 2015: 183). In this sense, non-state action could enable a transformation of the climate change response in its own right – away from a purely top-down state-led approach, to one where any type of non-state actor, including individuals, could act on their climate responsibilities (Turnheim et al. 2018: 3). This is important, because the 'political reality' of climate change is that actors will not automatically comply with their moral responsibilities (Caney 2014: 134).

It is important to remember that climate change actors operate in social, political and economic contexts. Action on climate change is often influenced by these contexts, for example by the political context of Republican power brokers in the

United States, which has affected the US role in the Paris Agreement, or the economic context of increasingly cost-effective green energy, which has affected the Indian government's decision to scale back plans for new coal-fired power plants in favour of solar energy farms (Dietzel 2019: 101). According to Caney (2014: 135) it is possible to structure these social, economic and political contexts 'in ways which induce agents to comply with climate responsibilities'. Caney explains that restructuring can include enforcing compliance through policy measures, incentivizing actors by offering rewards for their action, creating norms which encourage compliance by making non-compliance seem unacceptable, undermining resistance to compliance through accurate media representation of climate change, and by using civil disobedience to encourage governments to act (Caney 2014: 136–8). The capability to change the context of the response is imperative if we are to implement the wide-scale system change required to tackle climate change in a just manner. There will need to be changes in behaviour at every scale, and these changes will not be possible unless the context of the climate change is shifted substantially.

There are several reasons to be optimistic about the role of non-state actors in restructuring the context to make space for a just response to climate change. For one, their participation has allowed for an independent and creative pursuit of climate change responses that were not possible before these actors emerged. For example, when Donald Trump withdrew from the Paris Agreement, a strong national response emerged: a coalition of non-state actors who call themselves 'We Are Still In' (2019). This collection of 3783 (and counting) climate leaders, which includes mayors, county executives, governors, tribal leaders, college and university leaders, businesses, faith groups and investors, aims to 'support climate action to meet the Paris Agreement' (We Are Still In 2019). Such a scale of climate action has never been seen before in the United States: to date, 'We Are Still In' is the largest cross-section of the American economy yet assembled in pursuit of climate action (We Are Still In 2019). The emergence of global non-state action has created a context where initiatives such as 'We Are Still In' are taken seriously and able to play a significant role in the response to climate change, even without national support.

Not only is there scope for creative and independent action, there is also scope for greater ambition. Unconstrained by the consensual decision making found in international treaty-making processes, 'there are virtually no limits on what non-state actors can do to respond to climate change' (Bulkeley et al. 2015: 164). They can push the envelope of what is possible by 'actively seeking out gaps in the response to climate change and attempting to fill them' (Hoffman 2011: 78). Many non-state initiatives include targets and timetables that go far beyond those agreed by states (Bulkeley and Newell 2010: 67), and they have been known to push state actors directly during negotiations. At COP21 in Paris, for example, members of the Climate Action Network (a worldwide network of over 1300 non-governmental organizations) called on states to improve their pledges ahead of 2020 (Earth Negotiations Bulletin 2015: 13). Furthermore, the creativity, ambition and innovation of non-state responses has the potential to catalyse significant change – for example changes to our growth-based global economic system, or to meat- and dairy-heavy food production systems, or even to our fossil fuel-reliant energy systems (Ostrom 2009: 4). Such change is necessary after years of incremental movement by the state-led global climate regime. Without non-state actors, emissions will not be cut enough to avert the worst effects of climate change, and funds for adaptation will continue to be inadequate (Dietzel 2019: 212).

Conclusion

Non-state actors are reshaping the context of the climate change response in a way that might allow for a more just world by lowering global greenhouse gas emissions, raising adaptation finance, enabling responsible actors to act and creating space for new and innovative responses to climate change. Climate justice scholars should take note of this contribution and explore the role of non-state actors in addition to, or alongside, international responses of states. They should be particularly interested in how non-state climate change responses compare to activities of states, individually and collectively, in terms of the just distribution of burdens, allocation of responsibility, protection of humans and providing for procedural justice. In exploring these new avenues, climate justice scholars can stay on top of the most recent developments in the climate change responses and be able to offer up-to-the-minute policy advice.

Cosmopolitan climate justice scholars can play a role in developing new methods of assessment to capture the contribution of non-state actors more precisely. A cosmopolitan agenda for climate justice critiques states as exclusive holders of climate-related responsibility and vehicles for climate justice. Individuals and other non-state actors are also responsible and able to contribute to realizing climate justice. Increasing numbers of non-state actors, ranging from business to cities and everything in between, are increasingly involved with climate change – not least because states have failed to do enough – and billions of individuals are capable of action to mitigate their climate-related behaviours. Indeed, the post-Paris Agreement climate regime treats these actors as key parts of effective responses to climate change. Adopting a cosmopolitan perspective can help scholars, and indeed practitioners, to comprehend more of the practical and normative complexities of climate change. A cosmopolitan viewpoint fully discerns the responsibilities of individuals and other non-state actors for climate change mitigation, adaptation and assistance. It is through this discernment that researchers might contribute to the eventual realization of climate justice in the future.

Acknowledgements

The authors have drawn from their Edinburgh University Press books (Dietzel 2019, Harris 2010, 2016) and works cited therein.

References

Abbott, Kenneth W. (2012), 'The transnational regime complex for climate change', *Environment and Planning C: Government and Policy*, 30(4), 571–90.

Archibugi, Daniele (2004), 'Cosmopolitan guidelines for humanitarian intervention', *Alternatives*, 29(1), 1–22.

Asian Cities Climate Change Resilience Network (2019a), *About Us*, accessed 19 April 2019 at www.acccrn.net/about-acccrn.

Asian Cities Climate Change Resilience Network (2019b), *Initiatives Map*, accessed 19 April 2019 at www.acccrn.net/map.

Asian Cities Climate Change Resilience Network (2019c), *Members*, accessed 19 April 2019 at www.acccrn.net/members.

Barry, John (1999), 'Statism and nationalism: a cosmopolitan critique', in Ian Shapiro and Lea Brilmayer (eds), *Global Justice*, New York: New York University Press, pp. 12–66.

Beitz, Charles (1999), *Political Theory and International Relations*, Princeton, NJ: Princeton University Press.

Brock, Gillian (2009), *Global Justice: A Cosmopolitan Account*, Oxford: Oxford University Press.

Bulkeley, Harriet and Peter Newell (2010), *Governing Climate Change*, London: Routledge.

Bulkeley, Harriet, Liliana B. Andonova, Michele M. Betsill, Daniel Compagnon, Thomas Hale, Matthew J. Hoffmann, Peter Newell et al. (2015), *Transnational Climate Change Governance*, Cambridge: Cambridge University Press.

Caney, Simon (2014), 'Two kinds of climate justice: avoiding harm and sharing burdens', *The Journal of Political Philosophy*, 22(2), 125–49.

Climate Group (2012), *Lighting the Green Revolution*, accessed 22 April 2019 at www.theclimategroup.org/_assets/files/LED_report_web1%283%29.pdf.

Climate Group (2019a), *Our Achievements*, accessed 22 April 2019 at www.theclimategroup.org/our-achievements.

Climate Group (2019b), *Who We Are*, accessed 22 April 2019 at www.theclimategroup.org/who-we-are/about-us/.

Dietzel, Alix (2017), 'The Paris Agreement: protecting the human right to health?', *Global Policy*, 8(3), 313–21.

Dietzel, Alix (2019), *Climate Justice and Climate Governance: Bridging Theory and Practice*, Edinburgh: Edinburgh University Press.

Dower, Nigel (2007), *World Ethics*, Edinburgh: Edinburgh University Press.

Earth Negotiations Bulletin (2015), *Summary of the Paris Climate Change Conference*, accessed 19 April 2019 at enb.iisd.org/climate/cop21/enb.

Gibney, Matthew J. (2015), 'Refugees and justice between states', *European Journal of Political Theory*, 14(4), 448–63.

Guterres, António (2019), 'Address to Fijian Parliament', accessed 17 May 2019 at https://www.un.org/sg/en/content/sg/speeches/2019-05-15/address-fijian-parliament.

Hale, Thomas (2016), '"All hands on deck": the Paris Agreement and nonstate climate action', *Global Environmental Politics*, 16(3), 12–22.

Harris, Paul G. (2008), 'Climate change and global citizenship', *Law and Policy*, 30(4), 481–501.

Harris, Paul G. (2010), *World Ethics and Climate Change: From International to Global Justice*, Edinburgh: Edinburgh University Press.

Harris, Paul G. (ed.) (2011a), *Ethics and Global Environmental Policy: Cosmopolitan Conceptions of Climate Change*, Cheltenham, UK and Northampton, MA, USA: Edward Elgar Publishing.

Harris, Paul G. (2011b), 'Introduction: cosmopolitanism and climate change and policy', in *Ethics and Global Environmental Policy: Cosmopolitan Conceptions of Climate Change*, Cheltenham, UK and Northampton, MA, USA: Edward Elgar Publishing, pp. 1–19.

Harris, Paul G. (2016), *Global Ethics and Climate Change*, Edinburgh: Edinburgh University Press.

Held, David (2005), 'Principles of cosmopolitan order', in Gillian Brock and Harry Brighouse (eds), *The Political Philosophy of Cosmopolitanism*, Cambridge: Cambridge University Press, pp. 10–27.

Heyward, Clare and Daniel Roser (eds) (2016), *Climate Justice in a Non-Ideal World*, Oxford: Oxford University Press.

Hoffman, Matthew J. (2011), *Climate Change Governance at the Crossroads: Experimenting with a Global Response after Kyoto*, Oxford: Oxford University Press.

Hsu, Angel, Andrew S. Moffat, Amy J. Weinfurter and Jason D. Schwartz (2015), 'Commentary: towards a new climate diplomacy', *Nature*, **5**(6), 501–503.

Lawrence, Peter (2014), *Justice for Future Generations: Climate Change and International Law*, Cheltenham, UK and Northampton, MA, USA: Edward Elgar Publishing.

Maltais, Aaron and Catriona McKinnon (ed.) (2015), *The Ethics of Climate Governance*, London: Rowman and Littlefield.

McKendry, Corina (2016), 'Cities and the challenge of multiscalar climate justice: climate governance and social equity in Chicago, Birmingham, and Vancouver', *Local Environment*, **21**(11), 1354–71.

Ooms, Gorik and Rachel Hammonds (2010), 'Taking up Daniel's challenge: the case for global health justice', *Health and Human Rights*, **12**(1), 196–218.

Ostrom, Eleanor (2009), 'A polycentric approach for coping with climate change', *World Bank Policy Research Working Paper*, no. WSP 5095, 1–54.

Oxfam (2017), *Extreme Carbon Inequality*, accessed 25 May 2019 at https://www.oxfam.org/en/research/extreme-carbon-inequality.

Pattberg, Phillip (2010), 'The role and relevance of networked climate governance', in Frank Biermann, Phillip Pattberg and Fariborz Zelli (eds), *Global Climate Governance Beyond 2012: Architecture, Agency, and Adaptation*, Cambridge: Cambridge University Press, pp. 146–64.

Pepper, Angie (2014), 'A feminist argument against statism: public and private in theories of global justice', *Journal of Global Ethics*, **10**(1), 56–70.

Pierre-Louis, Kendra (2018), 'Greenhouse gas emissions accelerate like a "speeding freight train" in 2018', *New York Times*, accessed 24 May 2019 at https://www.nytimes.com/2018/12/05/climate/greenhouse-gas-emissions-2018.html.

Sander, Chan, Harro van Asselt, Thomas Hale, Kenneth W. Abbott, Marianne Beisheim, Matthew Hoffmann, Brendan Guy et al. (2015), 'Reinvigorating international climate policy: a comprehensive framework for effective nonstate action', *Global Policy*, **6**(4), 466–73.

Scarborough, Peter, Paul N. Appleby, Anja Mizdrak, Adam D.M. Briggs, Ruth C. Travis, Kathryn E. Bradbury and Timothy J. Key (2014), 'Dietary greenhouse gas emissions of meat-eaters, fish-eaters, vegetarians and vegans in the UK', *Climatic Change*, **125**(2), 179–92.

Seyfang, Gill (2005), 'Shopping for sustainability: can sustainable consumption promote ecological citizenship?', *Environmental Politics*, **14**(2), 290–306.

Stevenson, Hayley and John S. Dryzek (2014), *Democratizing Global Climate Governance*, Cambridge: Cambridge University Press.

Turnheim, Bruno Jacob, Paula Kivimaa and Frans Berkhout (eds) (2018), *Innovating Climate Governance*, Cambridge: Cambridge University Press.

UNFCCC (2015), *Adoption of the Paris Agreement*, accessed 22 April 2019 at unfccc.int/documentation/documents/advanced_search/items/6911.php?priref=600008831.

Vanderheiden, Steve (2008), *Atmospheric Justice*, Oxford: Oxford University Press.

Vermeulen, Sonja J., Bruce M. Campbell and John S.I. Ingram (2012), 'Climate change and food systems', *Annual Review of Environment and Resources*, **37**, 195–222.

We Are Still In (2019), *Declaration*, accessed 22 May 2019 at https://www.wearestillin.com/we-are-still-declaration.

8 Social justice and ecological consciousness: pathways to climate justice

James S. Mastaler

Climate change is a structural problem with pernicious social and environmental consequences for the world's most vulnerable populations and fragile ecosystems. Structural problems, understood broadly as those emerging from the entrenched social patterns giving shape to both individual and public life, do not easily change. Realizing climate justice within the context of possible runaway climate change requires a critical review of what has gone wrong and why, as well as a response for action. This chapter offers a brief assessment of climate change as it relates to social justice and ecological consciousness. Climate change offers a challenge to longstanding ways of thinking about age-old fundamental questions regarding what it means to be human in relation to others and the world. Social justice is both an interpretive lens and a process of mobilization for the shared flourishing of individuals, groups and the whole of society. It is grounded in the principles of basic fairness, equity and participatory forms of decision-making. It facilitates transformational change across social and institutional structures from the grass-roots upward. Ecological consciousness is a way of seeing oneself in relation to others that can inform, repair and fortify new ways of thinking and being in light of contemporary environmental challenges.

A global movement toward climate justice can work to eliminate the most detrimental forms of global poverty, social inequality and environmental degradation. It ought to prioritize the most pressing tasks necessary for realizing a basic agenda for climate justice. A basic agenda for climate justice, as a response to pressing social justice concerns, is one that (1) mitigates current and future greenhouse gas emissions; (2) develops governance structures that work at both the local and global levels of society; (3) cultivates shared flourishing via poverty alleviation and increased community resilience; and (4) expands support for individuals and communities already experiencing the worst effects of climate change. Each of these tasks present research trajectories ripe for expansion and further investigation. As important as they are for instituting climate justice in the near future, however, they are not nearly extensive enough to carry us much further beyond where we ought to have been already. A vision of climate justice for the long-haul needs to be more holistic, wide-ranging and much bolder. We need to do more.

A more robust vision of climate justice relies on an expanded understanding of ecological consciousness. It is one that situates the human person in a mode of thinking about the individual self in terms of its network of relations with others and with the world. A vision of climate justice with ecological consciousness as its cornerstone also (1) advances cultures of care and compassion, and (2) establishes disciplines of intentional reflection and continuous examination that serve the individual's pursuit of a life well lived and that permeate into the individual's contributions toward public life and the common good. Future research in these two areas could further illuminate, expand and enhance the long-term work of climate justice.

Social justice and climate change

Climate change is a matter of social justice (Moss 2009). Human-induced climate change is the result of greenhouse gases emitted into the atmosphere at levels that began to soar dramatically as the Industrial Revolution transformed commerce and industry in Western economies. It is a scientific, economic and political phenomenon with social and environmental consequences that are common to all but that burden the world's most vulnerable populations disproportionately (Brainard et al. 2009). The collective actions of individuals, of businesses and corporations, and of governments and multinational conglomerates, have each contributed to the ongoing exacerbation of climate change. So far, the social systems, structures and institutions governing human civilizations have been unable to adequately mitigate climate change despite its dire consequences.

The failure to act on climate change represents massive negligence on the part of human civilizations to gather around a shared vision for change, to harness necessary resources, and to execute actionable steps forward. Great works of achievement, whether ancient or modern, require collective human effort and the coordination of knowledge and resources across whatever social systems, structures and institutions exist to facilitate them (Berry 1999). The pyramids at Giza and of the Maya across the Americas, the Great Wall of China, the European Enlightenment, the defeat of Nazism and fascism during World War II – including the rebuilding and eventual establishment of the United Nations not long afterward – and the skyscrapers and infrastructure that make living in modern cities possible, are examples of achievements that represent the power of vision and the marshalling of resources on a remarkable scale. How do we press ahead toward new great achievements and through the stagnation of climate inaction? Social justice offers an interpretive lens for understanding the problem in a way that promotes the active engagement of individuals and their communities with the systems and institutions governing our lives. Furthermore, social justice offers tools for changing the systems and institutions that have so far been unable to mitigate climate change, thereby alleviating some of its worst social and environmental consequences and thus furthering climate justice.

Climate change is a structural problem with social and environmental consequences

The structures that govern our lives together are the same structures through which we fail to mitigate climate change. The shortcomings of legal, political and governance institutions bear heavily upon the social sphere, contributing to vast disparities in access to food, clean water, medical care, education and other common goods necessary for basic health and wellness (Shue 2014). The severity of potential risks and impacts of climate change is greatest for least developed countries and vulnerable communities because such disparities create a limited ability for them to cope with all forms of environmental disasters, including hurricanes, flooding, heat waves and droughts (IPCC 2014). Moreover, modern agricultural, commercial and industrial practices strain the ecological stability of environments around the globe, and these degraded ecosystems are less able to bounce back from the type of natural disasters and extreme weather events forecasted for various climate change scenarios (Mintzer 1992). Ecosystem-dependent livelihoods support some of the world's most vulnerable working poor, which means these people are all too often devastated by increasing exposure risks to climate-change-related phenomena. These two factors – increasing vulnerability and sensitivity from longstanding legal and political shortcomings alongside the rising exposure risks from ongoing environmental degradation and climate change – demonstrate how current structural failures are likely to become increasingly compounded by climate change. The failure of the world's most powerful governments to curtail continued greenhouse gas emissions through the appropriate legal and political structures available to them also indicates a structural failure. It is a failure of the collective will to mitigate a problem with profound consequences – a problem generated by the systems and institutions of our own making and a problem that can only be remedied by working through the very channels that need to be changed.

Social justice offers an interpretive lens and a process for change management

A social justice framework offers a critically necessary interpretive lens for the most important justice-based issues emerging from climate change. Since those issues often revolve around implicit biases across a society's most powerful institutions, it is vital for justice that those in power consider whose interests are being served or, of equal importance, whose interests are not being served by those seated at decision-making tables. The world's most vulnerable populations are excessively sensitive and exposed to the risks presented by climate change, and they have little to no say regarding the policies that manage greenhouse gas emissions (IPCC 2014). Those who have the most to gain have disproportionate influence in decision-making. For example, large oil companies and governments that receive outsize profits from oil drilling, refining and consumption profit impressively from self-regulating economic markets as well as from unbinding, discordant legislation related to the regulation of emissions. Their disproportionate influence on the policies, procedures, systems and institutions governing us all represents a fundamental liability in the pursuit of basic fairness, social equity and environmental responsibility concerning the common good.

Social justice, as a correcting *interpretive lens*, seeks to rectify the liability of unrepresentative decision-making by acknowledging the role of participation in justice-making and transformational change. Social justice engages individuals and their communities in the collaborative work of transforming policies, procedures, systems and institutions so that they function more effectively for the common good of everyone in society (Elsbernd 1989). Social justice, furthermore, encourages greater representation for all members of society at each level of decision-making, especially for its most vulnerable members (Elsbernd and Bieringer 2002). As a *process of mobilization* for the common good, social justice embodies strategies of relationship building and community organizing that facilitate the increased social engagement of individuals (Frederking 2014). Social justice binds individuals to each other in solidarity and in participatory relationships of decision-making as well as to the principles of basic fairness, equity and shared flourishing.

Ecological consciousness and climate change

The development of ecological consciousness is a necessary part of the structural changes that will make climate justice possible. An ecological consciousness recognizes that life is formed in a community of relationships with other forms of life and that those relationships are the foundation of sustaining the larger community of life (Uhl 2013). The development of a distinctly modern ecological consciousness correlates with the rise of the hard and social sciences in the West. Consequently, it includes a way of thinking about what it means to be human in terms of not simply the biological web of life but also the web of relations that exist between the individual and society and between society and the ecosystems in which all beings are embedded. In its more developed iterations, ecological consciousness represents an integrated understanding of the self that is connected inseparably to other people, other species and the biomes of which we are each a part and upon which we all depend for our shared flourishing. A greater sense of ecological consciousness among individuals can make more people, as well as our social systems and institutions by extension and through sustained effort, more aware of and reactive to the underlying causes of climate change and its inherent injustices.

More importantly, ecological consciousness acts as a sustaining force that individuals and communities need in order to live out various ways of being in the world, which can help people to overcome the psychological barriers that make climate justice difficult to achieve (Garvey 2008). Since the phenomenon of climate change presents more than scientific, economic and political problems in need of technical solutions, our collective responses need to be more than simply technical themselves (Roser and Seidel 2017). Climate change presents a conceptual challenge to longstanding ways of thinking about age-old questions pertaining to what it means to be human in relation to others and the world.

The next subsection fleshes out a fuller description of this conceptual side of the problem. The subsequent subsection then suggests how the power and potential of more developed forms of ecological consciousness can help to inform, repair and fortify new ways of thinking that are inherently more appropriate to contemporary environmental and social contexts. Recognizing this side of the problem can better enable us to tackle climate change holistically and conscientiously and, therefore, take us a few steps further on a path toward climate justice.

Climate change presents challenges to longstanding ways of thinking

The conceptual challenge presented by climate change pertains to the fundamental way people imagine what it means to be human in the world. For much of Western civilization up to the Industrial Revolution, some general characteristics of prevailing worldviews are noteworthy. One is an image of the human person as a vulnerable being whose life is short-lived and fragile relative to the perceived mystery, power and vastness of nature. For much of human history, human populations have been relatively small, pocketed within agriculturally dependent communities and scattered across vast distances. Natural disasters and related phenomena, such as flooding or drought, have long manifested the power of nature and the frailty of human existence before it. Furthermore, without a modern understanding of how infectious diseases, such as plague, spread, illnesses and other natural occurrences presented as inexplicable phenomena steeped in mystery and superstition. In this context, as French (2010: 55) describes, '[t]he order of nature seemed to be a given, something whose existence and ongoing presence could be comfortably assumed. . .'.

This image of frail humanity before a mysterious, powerful and vast world mostly persisted in prevailing Western worldviews up through the 1700s, though there were some modifications to it beginning in the 1500s with the rise of the modern sciences. The sciences shifted prevailing worldviews away from a generally mysterious orientation toward a much more mechanistic conceptualization of nature (that is, a view of nature that is knowable and manipulable) (Merchant 1989). This parallels a surge in diverse streams of thought that begin to emphasize human agency and power alongside nature's power. Scientists like Francis Bacon and Sir Isaac Newton, philosophers like Rene Descartes and Immanuel Kant, and theologians like Martin Luther and John Calvin all contributed to this newer imagery of the human person. The rise of the modern environmental movement, specifically, and the emergence of human-induced climate change presents yet another new context for shaping the prevailing image of humanity in relation to nature.

Climate change flips the old imagery on its head completely; an image is emerging of nature as not merely mechanistic (knowable/manipulable) instead of mysterious, but also as frail instead of powerful, and also as limited instead of vast. This is set against an image of human domination – humanity's ability to mine the earth for energy and subdue the land to feed explosive population growth or our transformation of the very chemical composition of our oceans and our atmosphere – the pilfering of the global commons for economic gain, and the insatiable commercial appetite for natural resources to feed the gluttonous consumption of the world's most privileged elite. Modern imagery of humanity's presence as powerful and nature as frail has not, however, yielded a collective change to the way we live on the planet. These new images are still comingled with the older, deeply internalized stories we continue to tell ourselves about how we ought to function in society and live our lives within the world. O'Sullivan and Taylor (2004b: 8) have argued that the older, conceptual scaffolding gives rise to 'instrumental consciousness' through the way that it emphasizes the primarily instrumental nature of humanity's relationship to a world composed of natural resources for human use. The development of an ecological consciousness has arisen during the modern environmental movement, though it has also existed in various forms among some indigenous communities who have held an understanding of it for centuries; it offers an opportunity to integrate these newer images into a larger narrative that enables us to make better sense of how we ought to live in the world today. How people imagine themselves in the world can act as a counterweight to the ecological decline and injustice of human-induced climate change – if it changes the way people live in relation to the world.

Ecological consciousness offers to inform, repair and fortify new ways of thinking

Ecological consciousness, though it offers a worldview grounded in the hard and social sciences, must also be conversant with humanistic and/or religious values and traditions if it is to shape public life effectively. A strong ecological consciousness integrates various ways of knowing and being in the world that embody what O'Sullivan and Taylor (2004a: 3) have termed the 'ecological values' of, namely, 'connection, openness, generosity, appreciation, partnership, inquiry, dialogue and celebration'. Such values offer a pathway by which to inform emerging worldviews through scientific inquiry, interdisciplinary dialogue and cross-cultural partnerships. Also, they can repair persisting conceptualizations of the self and the world while fortifying new ways of thinking against more problematic values, specifically, the kind of rugged individualism, overly mechanistic image of nature, and instrumentalist relations that came to dominate major streams of thought in the 19th and 20th centuries.

The synergies of ecological consciousness and social justice are most promising when an *individual* develops a strong *personal* ecological consciousness on the one hand and when, on the other hand, *society* is bound in solidarity to *participatory relationships* of decision-making via the principles of basic fairness, equity and shared flourishing. Together, these drive the sustaining force that individuals and communities need to become empowered change-makers. In the context of climate change and its inherently ecological, social and structural injustices, the development and expansion of ecological consciousness is necessary to undergird not only the individual pursuit of a life well lived but also the individual's contributions to public life and the common good.

Essential tasks for basic climate justice

Social justice and ecological consciousness are able partners on a pathway toward climate justice, but they still require an action-based roadmap for the elimination of global poverty, social inequality and environmental degradation. Such a roadmap, as proposed here, contains essential tasks that individuals and communities can complete to actualize real change in social systems and institutions, resulting in direct outcomes for the world's most vulnerable populations. This is a programme for climate justice built on the premise that climate change is an inherent structural problem that presents fundamental challenges to longstanding ways of thinking. Any pathway toward climate justice needs to be both practical and ambitious in simultaneously addressing the twin pillars of social equity and environmental responsibility. A basic programme for climate justice includes the following essential tasks: mitigating current and future greenhouse gas emissions, developing governance structures that work at both the local and global levels of society, cultivating shared flourishing via poverty alleviation and increased community resilience, and expanding support for individuals and communities that are experiencing climate change now.

Mitigating current and future greenhouse gas emissions

International agreements, sustainable development projects, and all national, state and local programmes need to work to reduce current and future greenhouse gas emissions across all respective programmes, projects and initiatives. While it may not be feasible to do so in every situation, at all times, it must always be a point of consideration and effort. The international community has already failed in preventing climate change outright. Since it is a process already underway, climate change is now disrupting fragile economies, vulnerable infrastructures and frontline communities along low-lying island nations, coastal communities, river deltas, indigenous arctic villages as well as impacting the agriculture-dependent livelihoods of small-scale farmers across much of the Asian and the African continents (IPCC 2014). As unfortunate as this is, things will get much worse for these communities and for almost everyone else if the world continues to fumble forward under a business-as-usual scenario. An essential element of any climate justice programme is the requisite task of intercepting the still-preventable aspects of climate change (Mastaler 2019a). It is too late to prevent climate change entirely, but it is not too late to mitigate some of the worst-case scenarios that are still before us.

Developing governance structures that work at local and global levels of society

We need to develop governance structures that simply work better at facilitating collaborative action across both the local and global levels of society. Mitigation efforts to date have not been adequate, partly because the frameworks for social organization and cooperation are inadequate. A subsidiarity-based approach to governance applied across the local-to-global spectrum – one that relegates decisions and tasks to the smallest body capable of adequately adjudicating and accomplishing them but that moves them to larger bodies when necessary – is a helpful way to organize efforts in a way that matches the scale of a problem to the entity best equipped to handle it, and magnifies effective strategies up and across those spheres of activity that necessarily extend far beyond the local level (Grasso et al. 1995, Mastaler 2011, Colombo 2012). One serious limitation of this approach, however, is the lack of any kind of centralized authority tasked with or capable of prioritizing, managing and adjudicating the global environmental commons. What is needed is a kind of world-level environmental organization that is structured and empowered analogously to those institutions that govern the social and economic commons – namely, the World Health Organization (WHO), the International Monetary Fund (IMF) and the World Trade Organization (WTO). The moral foundation of a possible world environmental organization (WEO) could be grounded in an existing and strong ethical framework aimed squarely at human equity and ecological sustainability (Ogorzalek and Rabb 2018). The Earth Charter, which was drafted according to 'the most inclusive and participatory process ever associated with the creation of an international declaration' and has been endorsed by over 6000 organizations around the world, would offer legitimate ethical scaffolding for a world environmental organization (Earth Charter Initiative 2012). A WEO that balanced cooperative governance with enforcement authority on issues pertaining to the equitable use and sustainable management of natural resources would offer a centralized forum for managing some of the world's most complex problems.

Cultivating shared flourishing through poverty alleviation and community resilience

To improve a community's general resilience to climate change, they require inclusive and just sustainable development in the space between thresholds necessary for human flourishing and bio-physical planetary boundaries (Rockström et al. 2009). Therefore, we need to cultivate a basic level of shared flourishing, especially among the world's most vulnerable populations, specifically through programmes emphasizing education, equity, agricultural productivity and stability in rural areas, and infrastructure projects in dense urban areas (Mastaler 2019a). The way in which various individuals and communities contribute to the cultivation of this shared flourishing requires different actions based on social context (Dobson 2004). While some communities are impoverished and cannot presently provide necessary thresholds for human flourishing, others facilitate cultures of overconsumption and wastefulness, especially those whose fuel-heavy transportation habits and industrialized consumption patterns contribute the most per capita to greenhouse gas emissions. For shared flourishing to occur, everyone must contribute but what that contribution looks like depends on who one is and how one lives right now.

Since not everyone contributes equally to greenhouse gas emissions and some revel in wastefulness while many others barely survive in extreme poverty, the principles of basic fairness and environmental responsibility demand that we cultivate changes that actually encourage some to consume more resources while others must necessarily become more efficient in their consumption. Applied to the case of greenhouse gas emissions, the United Nations Framework Convention on Climate Change's mandate calls for 'common but differentiated responsibilities' as a part of any internationally agreed-upon climate solution (Stone 2004). While this concept has emerged from and has been applied within the context of nation-states, it should be extended to the way we think about relations among individuals and between social groups. Moreover, it should probably be applied beyond the topic of greenhouse gas emissions to include aspects of personal and public life that are central to the alleviation of poverty, increased resilience to climate change, environmental sustainability and ecological conservation.

Expanding support for individuals and communities experiencing climate change now

Vigorous expansions of structural support are necessary for the transition of communities disrupted by the effects of climate change and also those disrupted by the transition from earth-degrading to earth-enhancing paradigms (Harvey 2019). Two groups often pitted against each other share a common threat of displacement and a shared need for expanded support. This includes: (1) those physically displaced by the international community's failure to mitigate climate change, a varied global community that ranges from islanders in the Pacific to tribal communities in the Arctic; and (2) those economically displaced by the shift from a dirty economy to a clean economy, such as those employed by fossil fuel industries, including coal miners in Appalachia as well as those involved in tar sands extraction in Canada and hydraulic fracturing in North Dakota. Such workers struggle with displacement when they are unable to transition into the growing number of green economy jobs that are increasingly driving investment and economic development (Robinson 2018). A just and inclusive transition is more politically expedient if it is broad enough to include both vulnerable groups. Whether it is an indigenous arctic community whose ancient ways are at risk of being lost to history or a coal miner whose company has closed up shop and retooled in some other region or industry, this is an environmental, economic and international policy shift with severe local impacts on real individuals' lives. A just and inclusive transition toward a new climate paradigm is about justice for people and planet together, and it fails if victory for one is taken at the expense of the other or if victory for some comes at the expense of justice for all.

A vision of climate justice for the future

The programme for basic climate justice proposed already is ambitious, but it is not as bold as it rightly should be. An advanced or more progressive programme for climate justice extends beyond this baseline of necessary and immediate changes, by attending to tasks that are still inherently structural but also centred on actions that help individuals prepare for the future – actions that can empower, inspire and sustain an individual's contributions to collective change-making over the long haul. This is critical because even though we need to take many climate actions post haste, we also need new ways of living and being in relation to each other and the world that can be sustained farther into the future – and establishing those new ways is likely to be the great work of generations, one that this generation can only now begin but likely cannot finish. The vision for climate justice that follows facilitates the development of an ecological consciousness among individuals so that their understanding of self in relation to others and the world can participate constructively in the present task of building a climate justice movement now, while also laying the foundations for future movements. This vision seeks specifically to advance cultures of care and compassion and to establish disciplines of intentional reflection and continuous examination.

Advancing cultures of care and compassion

The climate justice movement embodies a concern for basic fairness and social equity in response to the environmental challenges associated with climate change. Advancing cultures of care and compassion for both people and planetary systems can advance this goal in two important ways. First, we ought to extend existing cultures of care and compassion to include concern for the whole human person in all of its complexity, including physical, intellectual, emotional and spiritual needs. All bodies need food, clean water and medical care; our minds need access to educational opportunities appropriate to our abilities; our hearts and souls need social support networks, enriching encounters in the natural world, and space for reflective contemplation (Mastaler 2019b). Until or unless such basic human needs are met it is unrealistic to expect most people will feel able to participate in the larger climate justice movement. Society can empower its members by working to ensure they have access to such necessary basic goods. Second, we ought to extend cultures of care and compassion to include concern for other species and the general well-being of Earth's biosystems. Do our communities preserve and conserve the land, water and air in ways that maintain biodiversity and functional, resilient ecosystems? While the earth and non-human forms of life may certainly matter intrinsically, surely human well-being and planetary well-being are woven together so inseparably that human flourishing is dependent upon ecological integrity. If climate justice is achievable, it will only be in a world where people and non-human forms of life persist and thrive together into the future. Advancing cultures of care and compassion by expanding concern for the whole person and for planetary systems would enable many more of us to join the climate justice movement at a time when all hands are needed on deck.

Establishing disciplines of intentional reflection and continuous examination

Incorporating the disciplines of intentional reflection and continuous examination can help to foster rewarding personal lives and responsible social transformations. When individuals reflect on the actions they can take to reduce their carbon footprint – such as taking public transportation, or when they consider ways to reduce their conspicuous consumption by taking in cultural programmes, perhaps, instead of shopping for unnecessary consumer products – their actions can simultaneously work to nurture personal well-being, a stronger social fabric and the common good (see Harris 2013: 171–96). Moreover, when individuals examine whether or how their decisions and actions help or hurt the collaborative mission to build a more just and verdant world, they are essentially engaging in a reflective process in which they hold themselves accountable to a vision of the world as they believe it can and ought to be. This kind of self-accountability can be expanded across a continuum of accountability that begins with individual effort and extends beyond it throughout one's various social networks and spheres of influence from religious communities to civic groups, places of employment and governing bodies. The work of building climate justice begins by holding ourselves accountable to the way we live in relation to others and the world, and it grows as we amplify that accountability at work, at houses of worship, in business, or in government. Acknowledging the structural nature of climate change does not reduce the primacy of personal action or our responsibility for the social structures in which we are embedded and participate; neither does it demand that individual action shoulder all the burden. Rather, it recognizes the complexity of climate change and offers a way to engage with it. Transformational change pursued intentionally and held to account, whether from the individual pursuit of a life well lived or by the individual's contributions to public life and the common good, is critical for climate justice.

Conclusion

The collective negligence on climate change has amounted to a failure of justice concerning the world's most vulnerable populations and fragile ecosystems. Since climate change is a structural problem, a social justice response is necessary. The concept of social justice offers an interpretive lens and a process that is effective across social and institutional structures at every level of society. Additionally, since climate change presents a challenge to the longstanding ways of thinking about fundamental questions regarding what it means to be human in relation to others and the world, emerging forms of ecological consciousness are vital in shaping new ways of thinking. If climate justice is possible, it must work to eliminate the most pernicious forms of global poverty, social inequality and environmental degradation. The way to climate justice prioritizes the most immediate, essential tasks of a basic agenda and then subsequently ramps up toward more advanced approaches as it becomes feasible.

This chapter has both outlined a basic agenda for climate justice and has identified aspects of a more advanced or progressive approach. Each of the proposed tasks in either track plays an essential role in creating pathways to climate justice and any of them could serve as a direction for research. But, a holistic vision of climate justice that is grounded in the needs and concerns of both people and planet expands social justice and an emerging ecological consciousness into the cornerstones of the climate justice movement. In its most forward-looking iterations, this vision of climate justice advances cultures of care and compassion while establishing ongoing practices of intentional reflection and examination across a continuum of accountability spanning individual action and global collaboration. Social justice and ecological consciousness offer fresh approaches and undergird valuable ways of thinking about what it means to be human in relation to others and the world; both are necessary components of the short- and long-term structural changes that will determine whether climate justice is possible.

References

Berry, Thomas (1999), *The Great Work: Our Way into the Future*, New York: Random House.

Brainard, Lael, Abigail Jones and Nigel Purvis (eds) (2009), *Climate Change and Global Poverty: A Billion Lives in the Balance*, Washington, DC: Brookings Institution Press.

Colombo, Alessandro (ed.) (2012), *Subsidiarity Governance: Theoretical and Empirical Models*, New York: Palgrave Macmillan.

Dobson, Andrew (ed.) (2004), *Fairness and Futurity: Essays on Environmental Sustainability and Social Justice*, Oxford, UK and New York, NY, USA: Oxford University Press.

Earth Charter Initiative (2012), *What is the Earth Charter?*, accessed 5 April 2019 at www.earthcharter. org/discover/what-is-the-earth-charter/.

Elsbernd, Mary (1989), *A Theology of Peacemaking: A Vision, a Road, a Task*, London, UK and Lanham, MD, USA: University Press of America.

Elsbernd, Mary and Reimund Bieringer (2002), *When Love is Not Enough: A Theo-Ethic of Justice*, Collegeville, MN: Liturgical Press.

Frederking, Lauretta Conklin (2014), *Reconstructing Social Justice*, New York: Routledge.

French, William C. (2010), 'With radical amazement: ecology and the recovery of creation', in David Albertson and Cabel King (eds), *Without Nature? A New Condition for Theology*, New York: Fordham University Press, pp. 54–79.

Garvey, James (2008), *The Ethics of Climate Change: Right and Wrong in a Warming World*, London, UK and New York, NY, USA: Continuum.

Grasso, Kenneth L., Gerard V. Bradley and Robert P. Hunt (eds) (1995), *Catholicism, Liberalism and Communitarianism: The Catholic Intellectual Tradition and the Moral Foundations of Democracy*, Lanham, MD: Rowman and Littlefield.

Harris, Paul G. (2013), *What's Wrong with Climate Politics and How to Fix it*, Cambridge: Polity.

Harvey, Samantha (2019), 'Leave no worker behind', in Laurie Mazur (ed.), *Resilience Matters: Strengthening Communities in an Era of Upheaval*, Washington, DC: Island Press, pp. 36–45.

IPCC (2014), *Climate Change 2014: Synthesis Report. Contribution of Working Groups I, II and III to the Fifth Assessment Report of the Intergovernmental Panel on Climate Change* [Core Writing Team, R.K. Pachauri and L.A. Meyer (eds)], Geneva, accessed 25 February 2019 at www.ipcc.ch/report/ar5/syr.

Mastaler, James S. (2011), 'A case study on climate change and its effects on the global poor', *Worldviews: Global Religions, Culture and Ecology*, 15(1), 65–87.

Mastaler, James S. (2019a), 'Social justice and environmental displacement', *Environmental Justice*, 12(1), 17–22.

Mastaler, James S. (2019b), *Woven Together: Faith and Justice for the Earth and the Poor*, Eugene, OR: Cascade Books.

Merchant, Carolyn (1989), *The Death of Nature: Women, Ecology and the Scientific Revolution*, New York: Harper and Row.

Mintzer, Irving M. (ed.) (1992), *Confronting Climate Change: Risks, Implications and Responses*, Cambridge: Cambridge University Press.

Moss, Jeremy (ed.) (2009), *Climate Change and Social Justice*, Melbourne: Melbourne University Press.

O'Sullivan, E.V. and M.M. Taylor (2004a), 'Conundrum, Challenge, and Choice', in E.V. O'Sullivan and M.M. Taylor (eds), *Learning Toward an Ecological Consciousness: Selected Transformative Practices*, New York: Pan Macmillan, pp. 1–4.

O'Sullivan, E.V. and M.M. Taylor (2004b), 'Glimpses of an ecological consciousness', in E.V. O'Sullivan and M.M. Taylor (eds), *Learning Toward an Ecological Consciousness: Selected Transformative Practices*, New York: Pan Macmillan, pp. 5–23.

Ogorzalek, K. and G.B. Rabb (2018), 'The case for a world environment organization', *Minding Nature*, 11(2), 26–35.

Robinson, Mary (2018), *Climate Justice: Hope, Resilience and the Fight for a Sustainable Future*, New York: Bloomsbury Publishing.

Rockström, Johan, Will Steffen, Kevin Noone, Asa Persson, F.Stuart Chapin III, Eric F. Lambin, Timothy M. Lenton et al. (2009), 'A safe operating space for humanity', *Nature*, 461(7263), 472–5.

Roser, Dominic and Christian Seidel (2017), *Climate Justice: An Introduction*, London, UK and New York, NY, USA: Routledge.

Shue, Henry (2014), *Climate Justice: Vulnerability and Protection*, Oxford: Oxford University Press.

Stone, Christopher D. (2004), 'Common but differentiated responsibilities in international law', *American Journal of International Law*, 98(2), 276–301.

Uhl, Christopher (2013), *Developing Ecological Consciousness: The End of Separation*, 2nd edn, Lanham, MD: Rowman and Littlefield Publishers.

9 Climate justice in practice: adapting democratic institutions for environmental citizenship

Giuseppe Pellegrini-Masini, Fausto Corvino and Alberto Pirni

Climate justice is a complex concept that compounds environmental, distributional (intergenerational and intragenerational) and procedural justice. In this chapter, we focus mainly on the intergenerational dimension of climate justice and its relevance to reshaping current conceptions of citizenship. In doing so, however, we will inevitably 'intersect' (Sovacool et al. 2017) with other dimensions of climate justice. A challenge for climate justice consists in the fact that uncoordinated actions risk overburdening individuals and creating unfair disadvantages between those who take an active role in preserving the environment and those who do not. Moreover, individual solutions to structural environmental problems, that is to say, environmental challenges that require the reform of collective norms and patterns, are likely to be ineffective in those cases in which both epistemological and moral obstacles drastically limit individuals' capacities to make sense of their own rights and responsibilities. Epistemological obstacles arise when individuals fail to appreciate the seriousness and urgency of detrimental phenomena such as climate change. Moral obstacles are those for which individuals, albeit being aware of collective risks, fail to trace out their individual responsibilities for contributing to causing those risks, and hence they lack motivation to act.

Additionally, different psychological mechanisms, which are based on subjective evaluations of lost and earned well-being, might lead even well-motivated individuals to postpone the performance of those actions that they perceive as just and necessary. The classic example, analysed below, is an individual who fully understands the range of climate change and recognizes a moral duty to act in order to prevent irreparable environmental damage, but who keeps on postponing the changes that might alter her lifestyle or routine. She does this because, with relation to a marginal 'wrongful action', the well-being lost by her is greater than the well-being which would be gained by the collectivity in case of non-performance of the same action.

Another challenge hinges on the distributive dimension of climate change, more specifically on the fair distribution of the burdens that just climate policies should allocate between current and future generations (Hale 2012). This will lead us to consider several climate citizenship policies that could be put in place to implement climate justice and to equitably share the burden of greenhouse gas emissions reduction among citizens while protecting the interests of future generations. We will discuss how climate citizenship policies can contribute to overcoming the obstacles that individuals encounter when they act in an institutional environment that holds a vision of climate justice reduced to voluntary behaviours and voluntary lifestyle adjustments, and we explore how government policies can ensure a fair representation of the interests of future generations.

The limits of individual morality and the need for institutional solutions to climate change

There are two main reasons why individual and uncoordinated solutions to climate change can be ineffective. The first reason is that climate change is a phenomenon in relation to which individuals tend toward 'moral corruption' (Gardiner 2011: 45). This corruption occurs when those involved in a complex moral problem profit from the existence of an intricate web of responsibilities to justify the preservation of the status quo. Consequently, halting environmental degradation cannot but require the introduction of institutional constraints to the pursuit of individual interest. The second reason is that even when single individuals manage to overcome moral corruption and are well disposed toward environmental preservation, the peculiar nature of climate change renders the cumulative costs of reversing it extremely high, while the marginal cost of a single polluting action can be relatively imperceptible (Diekmann and Preisendörfer 2003). This leads individuals to both first-order and second-order procrastination (Andreou 2007). Respectively, the first order of procrastination occurs when an individual agent postpones the actions that could help her in realizing what she perceives to be morally compelling (e.g. postponing the day to quit smoking). The second order of procrastination occurs when she postpones the implementation of the coercive solutions that might prevent postponing those same actions (e.g. postponing the day when a smoker asks his partner to adopt 'coercive measures' against him, such as hiding his cigarettes, in case he is unable to quit smoking alone).

The idea that the majority of actions against climate change are doomed to moral impasse has been explored by Gardiner (2011) through his metaphor of a 'perfect moral storm', which is supposed to illustrate the concurrence of three multilevel hostile classes of factors that reduce the chance of finding a solution to the most compelling collective problem of our age. The first class of factors is global in nature and concerns the existing asymmetry of power at the international level reverberating negatively on incentive schemes for the transition to less-polluting forms of production. Accordingly, the wealthier and more powerful economic agents are those that can benefit the most from 'business as usual', while the poorer and more vulnerable counterparts of the global economic system are those that suffer the most from climate change due to geography and because they are more reliant on agricultural production and lack the economic means for adapting to eco-systemic changes. This means that self-interested and powerful economic agents have economic reasons to refrain from reformist actions and instead pass on the negative costs of their productive activities to the poorest of the world (Gardiner 2011: 24–32).

The second class of factors that hinder the implementation of effective solutions to environmental challenges is intergenerational: externalities of climate change can not only be passed along through national borders but also across generations. In other words, the economic advantages of carbon dioxide can be reaped right now and the costs can be deferred in time. To this, we should also add that the present generation has to pay the costs of pollution from previous generations. These contingencies considerably reduce any individual's incentive to undertake virtuous behaviour (Gardiner 2011: 32–41). Consider, for example, the case in which a person rents an apartment and learns that she has to pay for the consumption of electricity in that apartment by the previous tenant. Assume also that this person, who is supposed to keep the apartment only for a couple of months, knows that she has no alternative but to pay past bills, otherwise the electric company will turn the lights off, while she can pass on with impunity the costs of her own electricity consumption to the tenant who will come next. It is evident that in a scenario like this, there is a strong incentive to perpetuate the chain of unjust actions rather than to interrupt it. Safeguarding the interests of future tenants would require the present tenant to absorb the costs of activities for which others have collected the majority of benefits. Accordingly, the only way to restore justice would be to introduce some institutional checks on the payment of bills and to make past consumers accountable for their consumptions. However, as in the case of climate change, many past free-riders will be dead. Consequently, it becomes necessary to agree upon new institutional mechanisms for redistributing historical negative externalities among present and future individuals.

Lastly, the theoretical storm identified by Gardiner is a consequence of the global and the inter-generational storms, which create serious problems to both moral and political theory when dealing with future generations (Gardiner 2011: 41–4). In our view, this theoretical storm should be split into two separate challenges. One is global and more easily solvable by relying on some empirical data, while the other one is intergenerational and requires, instead, some further theoretical elaboration. The global theoretical challenge consists in going beyond statist models of justice in order to incorporate the issue of the dispersed negative effects of climate change within a general discourse on justice. This can be achieved, for example, by relying on the empirical assumption that, given the globalization of economic systems, people living in different places take part in the production of a global cooperative surplus (Beitz 1979). Conversely, the intergenerational theoretical challenge is more complex, especially if we rely on models of justice, such as social contract theory (Arrhenius 2000), which are based on the assumption of mutual advantage. The question arises as to whether, as a matter of fairness, self-interested individuals should shoulder the costs of a transition toward more sustainable modes of production – which equates to a form of redistribution – in the interests of future generations from whom they cannot receive anything in exchange (Arrhenius 2000,

Pirni 2018a, 2018b, 2019). Here a sound theoretical solution will hinge upon the political significance and effectiveness of indirect cooperation, that is to say, cooperation between groups of agents that are misaligned in time, such that those in the middle can only receive from those who came before and redistribute toward those who will come after (Heath 2013). Accordingly, we cannot simply take for granted that there exist political reasons for keeping in place an intergenerational scheme of redistribution among people who can only reciprocate indirectly (Pellegrini-Masini et al. 2019).

In sum, the concurrence of global, inter-generational and theoretical obstacles makes it hard for the single individual to give due weight to climate change and to make sense of her own participation in the global and inter-generational chain of actions that lead to irreparable damages to the environment. Yet, even 'surviving' the perfect moral storm would not necessarily be enough for the individual unilaterally to change her behaviour against climate change. The person who overcomes both the epistemological and the moral obstacles related to climate change – and takes on board the moral need to reverse it – might still fail to act because that person is caught in an 'enduring preference loop' (Andreou 2007: 238). This happens when the marginal cost of an additional polluting action is so low, if taken singularly, and the aggregate costs of a comprehensive reform are so high for the person who is supposed to carry it out, that she might end up in a preference loop in which she is resolute in wanting a radical change but at any time she would prefer to slightly postpone the change in behaviour in order to enjoy the benefits of marginal pollution.

Consider, for example, how costly it would be for a person who happens to work far from home, in an area which is badly served by public transportation, to start using the bus instead of her car to get to her workplace so that she might reduce her individual emissions of carbon dioxide. Compare this with the small damage that a single additional car trip would bring to the environment. After all, the emissions produced by a single and short car trip would not make a substantial difference in the mathematics of global warming, while those emissions can have a substantial impact in terms of well-being for that specific individual on that specific day. These circumstances might trap the driver in a preference loop that leads her to continuously procrastinate over her change of routine. In a situation like this, one possible solution could be for this person to give herself a deadline to stop using the car, or for example to ask her partner to lock the garage door starting from a specific day, so that she will be forced to do what she understands to be the best thing for the environment and for the collectivity. Yet the same considerations just described might lead her to procrastinate over stopping procrastinating – that is, procrastinating in setting the deadline and in asking for help from her partner – thus falling prey to a second-order procrastination problem (Andreou 2007: 243–8).

If we extend this discourse to more complex individual dynamics leading to climate change or hindering the transition to renewable energy, we can understand why individual resolution (at least for many people) needs mechanisms of implementation which are externally enforceable to become effective. This poses a serious institutional issue. It calls into play a theoretical rethinking of the concept of citizenship in order to explain why institutions should push present individuals to take into account the interests of future generations, why procrastinating in doing so is unacceptable from a political point of view, and how they can effectively recognize political representation for future citizens. Below we analyse different legal and political arrangements for amending, in an intergenerational sense, the mechanisms of representative democracy.

Citizenship and climate justice

The concept of citizenship could be an important policy tool to help the individual to overcome the 'perfect moral storm' and keep up to her moral resolutions with regard to climate change and energy transition. In particular, a reformulation of the classic idea of citizenship so as to take into account the interests of future individuals has the potential to bound individual agency to deliver or protect public goods. Public goods, such as the environment and security, are by-products of social interaction that has been governed by formal rules, laws and informal personal and social norms (see, for example, Putnam et al. 1994). An attempt to stretch the boundaries of the social group that makes up the 'polis' can provide the present generation with the incentives to pass on public goods to future generations. In some sense, this would represent a top-down solution to the epistemological and moral challenges posed by climate change. At the same time, it would offer the individual effective motivations to break the cycle of procrastination. Accordingly, national institutions that are shaped based on an upgraded conception of citizenships can sanction individual wavering with respect to more sustainable paths of production and consumption, thus making it rational to act now instead of postponing the costs of transition.

The specific field of study of environmental citizenship is an example of using citizenship to address environmental problems (Dobson 2003, Smith 1998). Harris (2008, 2016) convincingly addresses the problem of reconfiguring citizenship in the context of climate change. He does so by focusing on distributional processes affecting present generations and advocating for more robust distributional policies that should be enforced by governments through appropriate taxation of global wealthy individuals while preventing negative effects on the poor. He affirms that 'The needs of present generations should not be ignored for those of the future' and adds that 'the present does not trump the future' (Harris 2008: 494). This tension between protecting the interests of the present poor versus future generations appears to be an inevitable dilemma that modern societies confront, in both developed and developing economies, every time that there is a discussion surrounding the need for regulating some industrial sector that is considered to have a negative environmental impact while also often providing much-needed local jobs. This debate often neglects the evidence of the very limited magnitude of short-term competitiveness and job losses that may be produced by environmental regulations (Dechezleprêtre and Sato 2017).

The call of Harris to shift responsibility, at least to a larger extent, from nation-states to individuals is worth heeding, considering that consumption choices are the drivers of economies and affect carbon emissions (Harris 2008). It is appropriate to point to current deep inequalities in emissions (Harris 2008): the poorest 40 per cent of the world's population is responsible for only 10 per cent of the world's final energy use, and consequently of about that percentage of greenhouse gas emissions (Grubler et al. 2012). Many natural resources are finite and not renewable, implying a problem of inequality of distribution between past, present and future generations. Research has warned that we are on a path of depletion of natural resources (Meadows et al. 1973, Turner 2008), and this will have substantial consequences for future societies, even bringing into question their ability to satisfy basic needs like food, shelter and healthcare, not to mention the disproportionate impact on already deprived populations (Pennock et al. 2016). Even the countries with relatively ambitious carbon dioxide emissions reduction targets seem to be reluctant to place significant burdens on their citizens. For example, the European Union appears to have addressed the environmental impact of the lifestyles of its citizens mainly through communication campaigns aimed at motivating a shift in current behavioural patterns (Vihersalo 2017).

Melo-Escrihuela (2008) criticizes that advocates of environmental citizenship – the set of norms and behaviours of citizens with regard to the environment – have been focusing too much on individuals' responsibility and the potential contribution to sustainability that could be achieved through voluntary lifestyle and consumption changes, whereas collectivistic responsibility is often neglected. This appears to be plausible, particularly because pro-environmental campaigns and voluntary lifestyle changes would appear to be the most acceptable way of producing sustainable behaviours in a liberal capitalistic society. However, we know that this approach is not working. In fact, despite pro-environmental messages being widespread in the media and recognized by the majority of individuals, few people make substantial efforts to curb their greenhouse gas emissions, which translates into what has been called the 'attitude–behaviour' gap (Kollmuss and Agyeman 2002). It appears that the only solution that can provide a significant and fair shift in consumption may be a top-down policy approach that substantiates climate citizenship as a system of strict rights and duties, enshrined in law and opportunely enforced by authorities, aimed at reducing the impact of the current generations' activities on the environment and on future generations.

According to Faulks (2000: 13), 'citizenship is a membership status, which contains a package of rights, duties and obligations and which implies equality, justice and autonomy'. Reconceptualizing citizenship in terms of climate change will require refocusing Faulks' definition and extending it to future generations. Harris (2008) has already made a convincing argument toward applying a cosmopolitan view of justice and considering citizens as members of a global community. This has been long overdue, and the consequences in terms of dumping on citizens of underdeveloped economies the negative externalities of Western industrial processes are evident in phenomena caused by climate change, such as growing desertification and flooding, which are contributing to a rising number of climate refugees from underdeveloped economies (Ahmed 2017). A reconceptualization of citizenship in light of climate justice comes from the need for taking into account future generations precisely because some of the natural resources that we are using are non-renewable: their depletion will negatively affect future generations (Pennock et al. 2016, Bardi 2018).

Policies and parliamentary designs for climate citizenship

New policies can be implemented to realize climate citizenship, but to be effective they should be internationally agreed and implemented in the national legislation. Two complementary approaches at delivering climate citizenship could work in synergy to achieve intergenerational and intragenerational justice, following approaches developed in the energy justice debate (Sovacool and Dworkin 2015). These two approaches focus on the private sphere and the public sphere. In the latter case, we focus here on potential designs for political representation that would likely further climate justice.

Climate citizenship policies for the private sphere

Climate citizenship policies for the private sphere should address distributional intergenerational and intragenerational justice in everyday behaviours of citizens while ensuring environmental sustainability. Various types of policies fall under this rubric: income taxes, whose revenue is reinvested in environmental protection and compensation toward national and international recipients; greenhouse gas emissions taxes on products and services addressed to individuals above a specific income threshold, providing a compensation system for low-income citizens (the revenues could be spent on environmental compensation and protection projects at national and international levels); and a system of non-tradable personal carbon allowances (PCAs). The first two types are familiar because income taxes and environmental taxes on specific goods are in place in every country, although those taxes do not necessarily have a revenue destination aimed at financing environmental protection projects.

In contrast, PCAs are relatively new and have not yet been implemented. Jagers et al. (2010) found that perceived fairness, trust toward politicians and self-interest of respondents would influence the acceptance of this policy proposal. Personal carbon allowances attribute to every individual an annual amount of carbon emissions that could be 'spent' on any consumption of choice by the citizen (Wallace et al. 2010). This system would leave citizens free to allocate their carbon allowance and financial resources on the purchases of their choice, thereby not altering the preference-based demand of market economies but establishing a significant element of equality in attributing to all citizens a limit to their personal emissions (independently of income). An egalitarian element would be lost, at least in terms of distributional equality, if the PCAs were to be tradable, because trading of allowances would allow wealthy individuals to keep polluting as much as they wish, thereby discharging the burden of carbon-emissions reduction on poorer individuals and thus furthering the problem of climate injustice. Hyams (2009) favours tradable PCAs because he believes that wealthy individuals would at least pay a price and therefore might be discouraged from polluting. This observation is rather objectionable, both for the considerations earlier expressed and also because consumerism and capitalistic economies not tempered by effective redistributive policies or caps contribute to inequality and environmental injustice within states and globally (Mészáros 2001). As Wilkinson and Pickett put it (2010: 13), 'Consumerism is fed by status competition and intensified by inequality leading to an insatiable demand for ever higher standards. This is one of the most important obstacles to cutting greenhouse gas emissions'.

Designing representation for climate justice

Climate citizenship policies for the public sphere would deal with specific systems of representation for future generations. Citizenship deals with the rights and duties of citizens, including representation and rights of political agency. We focus here on the representation of future generations because climate citizenship, to be a tool of climate justice, needs to deal both with intergenerational and intragenerational equity, and therefore to extend citizenship to future generations. Systems of representation of future generations would be the most effective approach for dealing with the short-termism of current political systems (González-Ricoy and Gosseries 2016) – that is, the tendency of political systems to deal with present problems by neglecting the long-term effects of the policies that they generate. The political representation of future generations is problematic if it is the duty of elected parliaments. The interests of generations living in the far future will most likely be discounted in favour of the interests of present or immediately following generations. Financial sustainability and, to a lesser extent, environmental sustainability are regarded as problematic (Tremmel 2006b) in modern democracies. Both people's 'perception' and 'ideology' (Ederer et al. 2006: 130), and the public's tendency to consider environments stable over time, on one hand, and, on the other hand, the tendency of reducing complex choices to ideological issues, both contribute significantly to the problem. Further, the perception of immediate costs and benefits versus the uncertainty of future ones, and the expectations of future technological solutions for long-lasting problems (MacKenzie 2016b), appear to contribute to this problem, along with the immediate interests of political actors (MacKenzie 2016b).

MacKenzie (2016b: 31) has presented several institutional design solutions to solve this problem of short-termism. Nevertheless, the problems of 'authorization' and 'accountability', as well as the 'epistemic problem' (Karnein 2016), arise in various degrees in all of the proposed designs. There are issues of authority and accountability because future generations will not be capable of electing an institutional subject, nor will they be capable of holding it to account for its actions. Further, such institutions will face the difficult task of defining the interests of future generations, and they will have to foresee the future consequences of current policies.

The representation of interests of future citizens could be realized in several ways, but all are to some extent prone to a problem of misrepresentation, specifically the dishonest or at least mistaken representation of the interests at stake. Different systems, when combined, might reduce the problem of misrepresentation. In this respect, several constitutional designs are worthy of mention. The first group are constitutional laws aimed at protecting the environment. Tremmel (2006a) presents 21 examples of constitutional provisions, the oldest adopted by Italy in 1947, with most of them having been adopted since the 1970s. Forty-two constitutions with clauses that aim at environmental protection for future citizens were listed by Dirth (2018: 46), but only five have legally binding formulations, and Bhutan and Ecuador are the only states that reference the principle of 'intergenerational equity'.

Despite being useful, constitutional clauses may have a varying degree of effectiveness, not only because of their wording but also because of the specific provisions that are made regarding the institutions that should enforce them. Further, general provisions of constitutional implementation, and especially the opportunity to challenge laws as unconstitutional by individuals and/or state institutions, are a significant element that impacts on the application of constitutional provisions of environmental protection. Each state has its own approach to the enforcement of constitutional principles, and these are usually enacted by constitutional courts, which are called upon to settle controversies by other courts or institutions. It is safe to say that the larger the number of subjects entitled to appeal to a constitutional court, the more likely are the constitutional provisions to be enforced.

Having an ombudsman for future generations (Beckman and Uggla 2016, Jávor 2006) is another institutional design that has the potential to address intergenerational climate justice. An ombudsman was first implemented in Sweden in the early 19th century, and in recent years several states have created this institution for environmental protection and sustainability, and to represent future generations. Recent instances of ombudsmen protecting the rights of future generations occurred in Hungary between 2007 and 2012, and in Israel, where a Commission for Future Generations was created in 2001 and operated until 2006 (Beckman and Uggla 2016, Teschner 2013). An ombudsman may examine state policies and advise relevant institutions to amend them, or it may delay legislation and, in some cases, initiate litigation against the institutions whose actions or deliberations are considered against the interests that the ombudsman is intended to protect. In many democracies, both in common law and civil law systems (e.g. the UK and

Italy respectively), significant powers to abolish or amend laws are attributed to non-directly elected or unelected bodies. Thus, an ombudsman could hold such powers. A downside might be that the ombudsman's authority is concentrated in the hands of an individual person, or perhaps in a limited number of individuals (such as a commission), thereby limiting its ability to oversee all relevant legislative issues.

Another institutional design for protecting future generations that has been proposed, and in some cases implemented, regards the possibility of altering the structure of the parliament by creating quotas for members who are meant to represent future generations (Dirth 2018). Youth quotas are among the means that have been employed for this particular end, but, as duly noted by Dirth (2018) and MacKenzie (2016b), they do not appear to be successful since young generations will not extend much into the future consideration of the implications of policies, and because young people may be concerned with their current needs, much as adults are.

A better solution is proposed by Dobson (1996): a system in which the interests of generations to come will be protected by a 'constituency of proxy electors and proxy representatives', both of which would belong to an environmentally friendly 'lobby' or movement (Dobson 1996: 133). Dobson admits that the identity of the lobby will be difficult to establish. There is also a matter of democratic legitimacy: selecting a proxy electorate based on 'green' attitudes might attribute a higher status to a green lobby's members if they are allowed to vote for the common parliament *and* for the seats that will represent future generations. This would be at odds with the principles of universal suffrage, and eventually of procedural justice and formal equality of citizens, that is at the core of modern democracies (Illuzi 2014).

Another proposal for parliamentary representation of future generations could circumvent the issue of infringing the principles of procedural justice and formal equality: a 'general purpose randomly selected chamber' (Mackenzie 2016a: 282–3). This would entail the selection of citizens to a separate chamber through a process that represents 'each politically relevant group' (Mackenzie 2016a: 282). This chamber, which would be an addition to the main chamber of the parliament, would have 'soft powers', such as the ability to delay legislation, (albeit not indefinitely), to propose amendments to legislation and possibly the power of 'holding the government to account' (MacKenzie 2016a: 283). This proposal would provide the possibility for ordinary individuals not belonging to political lobbies to join the parliament, thus radically respecting the idea of procedural justice and formal equality. Problems might arise in polarized political communities wherever it is conceivable that randomly selected members of parliament would possibly coordinate their activities with political groups of the elected chamber that have the greatest degree of affinity with them. Furthermore, in countries with high levels of corruption, it is imaginable that ordinary citizens might offer their support in return for some short-term benefits.

The last parliamentary design that we want to point out is Ekeli's (2016) sub-majority rule model. Ekeli proposes that minorities in parliaments could be provided with privileges to advocate for future generations' interests when legislation might affect them significantly. These powers would be essentially twofold: the first would grant to a minority of no less than one-third of the chamber the right to delay contested legislation up to the next election, while the second would allow the same minority to demand a referendum to submit the final decision about a contested bill to voters. A criticism of this proposal might be that there is no assurance that experienced politicians would not make use of these rules to score political points against their adversaries, even if these powers are allowed only for cases of legislative proposals regarding future generations. In fact, a great number of bills would normally be considered to have some consequences for generations to come.

Conclusion

We have focused much of our discussion on institutional solutions for establishing climate citizenship through policies of political representation. Climate justice could be implemented through a number of different institutional designs that aim to represent future generations. A mix of different institutional designs might be best for overcoming the shortcomings of any single design. Considering that procedural justice and formal equality might be regarded as core principles of modern democracies, it is arguable that any institutional design that aims to fulfil climate justice should be chosen only when upholding these principles.

We have discussed some of the challenges of enacting climate citizenship policies. We have highlighted some of the moral and psychological mechanisms that make it implausible to assume that voluntary behaviours, even when sustained by public campaigns for strengthening pro-environmental attitudes, will generate the level of climate citizenship needed to address climate change effectively. Only climate citizenship policies affecting both the public and private sphere of citizens' lives, internationally subscribed and implemented in national legislation, can create and enforce the new rights and duties that are needed to face climate change. Reconceptualizing citizenship alongside the urgency of tackling climate change requires realizing a kind of climate justice that would thread between individual freedoms and responsibilities at intragenerational and intergenerational levels. It also requires a substantial rethinking of a global economic system centred on accumulation and consumerism to the detriment of equality and distributional justice of financial and environmental goods.

While the policies discussed here do not comprise all of the possible institutional solutions to achieve climate-just citizenship, we believe that further developing this discussion by involving academics, policymakers, climate activists and all other relevant stakeholders is urgent. Realizing climate justice will require pushing for policy reforms that, along with technological innovation and green industrial policies and investments, accelerate the pace of the transition toward a carbon-neutral future. Neglecting the citizenship dimension of climate change would signify trying to borrow more time instead of averting severe environmental degradation and the injustices that come with it.

References

Ahmed, Bayes (2017), 'Who takes responsibility for the climate refugees?', *International Journal of Climate Change Strategies and Management*, **10**(1), 5–26.

Andreou, Chrisoula (2007), 'Environmental preservation and second-order procrastination', *Philosophy & Public Affairs*, **35**(3), 233–48.

Arrhenius, Gustaf (2000), 'Mutual advantage contractarianism and future generations', *Theoria*, **65**(1), 25–35.

Bardi, Ugo (2018), 'Peak oil, 20 years later: failed prediction or useful insight?', *Energy Research and Social Science*, **48**(March), 257–61.

Beckman, Ludvig and Fredrik Uggla (2016), 'An ombudsman for future generations – legitimate and effective?', in I. González-Ricoy and A. Gosseries (eds), *Institutions for Future Generations*, Oxford: Oxford University Press, pp. 117–34.

Beitz, Charles R. (1979), *Political Theory and International Relations*, Princeton, NJ: Princeton University Press.

Dechezleprêtre, Antoine and Misato Sato (2017), 'The impacts of environmental regulations on competitiveness', *Review of Environmental Economics and Policy*, **11**(2), 183–206.

Diekmann, Andreas and Peter Preisendörfer (2003), 'Green and Greenback', Rationality and Society, **15**(4), 441–72.

Dirth, Elizabeth (2018), *A Global Review of the Implementa on of Intergenerational Equity*, accessed 15 February 2019 at https://dspace.library.uu.nl/bitstream/handle/1874/361333/E_Dirth_5786851_Thesis.pdf?sequence=2&isAllowed=y.

Dobson, Andrew (1996), 'Representative democracy and the environment', in W.M. Lafferty and J. Meadowcroft (eds), *Democracy and the Environment: Problems and Prospects*, Cheltenham, UK and Brookfield, VT, USA: Edward Elgar Publishing, pp. 124–39.

Dobson, Andrew (2003), *Citizenship and the Environment*, Oxford: Oxford University Press.

Ederer, Peer, Philipp Schuller and Stephan Willm (2006), 'The economic sustainability indicator', in J.C. Tremmel (ed.), *Handbook of Intergenerational Justice*, Cheltenham, UK and Northampton, MA, USA: Edward Elgar Publishing, pp. 129–47.

Ekeli, Kristian Skagen (2016), 'Electoral design, sub-majority rules and representation for future generations', in I. González-Ricoy and A. Gosseries (eds), *Institutions for Future Generations*, Oxford: Oxford University Press, pp. 214–27.

Faulks, K. (2000), *Citizenship*, New York: Routledge.

Gardiner, Stephen M. (2011), *A Perfect Moral Storm: The Ethical Tragedy of Climate Change*, New York: Oxford University Press.

González-Ricoy, Iñigo and Axel Gosseries (2016), 'Designing institutions for future generations: an introduction', in I. González-Ricoy and A. Gosseries (eds), *Institutions for Future Generations*, Oxford: Oxford University Press, pp. 3–23.

Grubler, Arnulf, Thomas B. Johansson, Luis Mundaca, Nebojsa Nakicenovic, Shonali Pachauri, Keywan Riahi, Hans-Holger Rogner et al. (2012), 'Energy primer', in T.B. Johansson, A. Patwardhan, N. Nakicenovic and L. Gomez-Echeverri (eds), *Global Energy Assessment*, Cambridge: Cambridge University Press, pp. 99–150.

Hale, Benjamin (2012), 'Climate justice', in S.G. Philander (ed.), *Encyclopedia of Global Warming and Climate Change*, 2nd edn, Thousand Oaks, CA: SAGE Publications, pp. 359–61.

Harris, Paul G. (2008), 'Climate change and global citizenship', *Law & Policy*, **30**(4), 481–501.

Harris, Paul G. (2016), *Global Ethics and Climate Change*, Edinburgh: Edinburgh University Press.

Heath, Joseph (2013), 'The structure of intergenerational cooperation', *Philosophy & Public Affairs*, **41**(1), 31–66.

Hyams, Keith (2009), 'A just response to climate change: personal carbon allowances and the normal-functioning approach', *Journal of Social Philosophy*, **40**(2), 237–56.

Illuzi, Micahel J. (2014), 'Equality', in Michael T. Gibbons (ed.), *The Encyclopedia of Political Thought*, Chichester: John Wiley & Sons.

Jagers, Sverker C., Åsa Löfgren and Johannes Stripple (2010), 'Attitudes to personal carbon allowances: political trust, fairness and ideology', *Climate Policy*, **10**(4), 410–31.

Jávor, Benedek (2006), 'Institutional protection of succeeding generations – Ombudsman for Future Generations in Hungary', in J.C. Tremmel (ed.), *Handbook of Intergenerational Justice*, Cheltenham, UK and Northampton, MA, USA: Edward Elgar Publishing, pp. 282–98.

Karnein, Anja (2016), 'Can we represent future generations?', in I. González-Ricoy and A. Gosseries (eds), *Institutions For Future Generations*, Oxford: Oxford University Press, pp. 83–97.

Kollmuss, Anja and Julian Agyeman (2002), 'Mind the gap: why do people behave environmentally and what are the barriers to pro-environmental behaviour?', *Environmental Education Research*, **8**(3), 239–60.

MacKenzie, Michael K. (2016a), 'A general-purpose, randomly selected chamber', in I. González-Ricoy and A. Gosseries (eds), *Institutions For Future Generations*, Oxford: Oxford University Press, pp. 282–98.

MacKenzie, Michael K. (2016b), 'Institutional design and sources of short-termism', in I. González-Ricoy and A. Gosseries (eds), *Institutions for Future Generations*, Oxford: Oxford University Press, pp. 24–48.

Meadows, Donella H., Dennis L. Meadows, Jorgen Randers and William W. Behren (1973), 'The Limits to Growth: a report for the Club of Rome's project on the predicament of mankind', *Demography*, **10**(2), 295.

Melo-Escrihuela, Carme (2008), 'Promoting ecological citizenship: rights, duties and political agency', *Acme*, **7**(2), 113–34.

Mészáros, István (2001), 'The challenge of sustainable development and the culture of substantive equality', *New York*, **53**(June), 1–9.

Pellegrini-Masini, Giuseppe, Fausto Corvino and Lars Löfquist (in press), 'Energy justice and intergenerational ethics: theoretical perspectives and institutional designs', in Gunter Bombaerts, Kirsten Jenkins and Yekeen Sanusi (eds), *Energy Justice Without Borders*, New York: Springer.

Pennock, Michael, Blake Poland and Trevor Hancock (2016), 'Resource depletion, peak oil, and public health: planning for a slow growth future', in Isaac Luginaah and Rachel Bezner Kerr (eds), *Geographies of Health and Development*, Abingdon: Ashgate Publishing, pp. 177–96.

Pirni, Alberto (2018a), 'Intergenerational dwelling: the right to transform, the duty to preserve', *Studia Culturae*, **37**, 25–37.

Pirni, Alberto (2018b), *La sfida della convivenza. Per un'etica interculturale*, Pisa: Edizioni ETS.

Pirni, Alberto (2019), 'Overcoming the motivational gap: A preliminary path to rethinking intergenerational justice', *Human Affairs*, **29**(3), 286–296.

Putnam, Robert D., Robert Leonardi and Raffaella Y. Nanetti (1994), *Making Democracy Work: Civic Traditions in Modern Italy*, Princeton, NJ: Princeton University Press.

Smith, Mark J. (1998), *Ecologism: Towards Ecological Citizenship*, Minneapolis, MN: University of Minnesota Press.

Sovacool, Benjamin K. and Michael H. Dworkin (2015), 'Energy justice: conceptual insights and practical applications', *Applied Energy*, **142**(Supplement C), 435–44.

Sovacool, Benjamin K., Matthew Burke, Lucy Baker, Chaitanya Kumar Kotikalapudi and Holle Wlokas

(2017), 'New frontiers and conceptual frameworks for energy justice', *Energy Policy*, 105(March), 677–91.

Teschner, Naama (2013), *Official Bodies That Deal With the Needs of Future Generations and Sustainable Development – Comparative Review*, accessed 15 February 2019 https://www.knesset.gov.il/mmm/data/pdf/me03194.pdf.

Tremmel, Joerg Chet (2006a), 'Establishing intergenerational justice in national constitutions', in J.C. Tremmel (ed.), *Handbook of Intergenerational Justice*, Cheltenham, UK and Northampton, MA, USA: Edward Elgar Publishing, pp. 187–214.

Tremmel, Joerg Chet (ed.) (2006b), *Handbook of Intergenerational Justice*, Cheltenham, UK and Northampton, MA, USA: Edward Elgar Publishing.

Turner, Graham M. (2008), 'A comparison of The Limits to Growth with 30 years of reality', *Global Environmental Change*, 18(3), 397–411.

Vihersalo, Mirja (2017), 'Climate citizenship in the European Union: environmental citizenship as an analytical concept', *Environmental Politics*, 26(2), 343–60.

Wallace, Katherine N., Katherine N. Irvine, Andrew J. Wright and Paul D. Fleming (2010), 'Public attitudes to personal carbon allowances: findings from a mixed-method study', *Climate Policy*, 10(4), 385–409.

Wilkinson, Richard and Kate Pickett (2010), *The Spirit Level – Why Greater Equality Makes Societies Stronger*, New York: Bloomsbury Press.

10 Climate refugees: realizing justice through existing institutions

Justin Donhauser

This chapter examines concrete proposals for addressing climate refugee issues. 'Climate refugees' is a term that describes migrants who are forced to flee their homelands and seek asylum due to experiencing irreparable loss or damage caused by events linked to global climate change. Such loss and damage includes, for example, loss or damage of property and resources from extreme storms, severe droughts and floods, sea-level rise and air and water contamination. Examples of climate refugees are Peruvian immigrants seeking asylum in the United States after their potable water and agricultural resources became poisonous due to heavy metals being released through rapid glacial melting (cf. Climate: observations 2011, Donhauser 2018).

Many sources document other people who have been forced to migrate due to events linked to climate change. And leading advisory organizations point out that the number of climate refugees will inevitably grow in coming years (see, for example, Climate Refugee (n.d.), Climate Change and Disasters 2015). Now unavoidable global temperature increases will continue to cause more food crop shortages, more severe and erratic weather events, and sea-level rise. These things alone will continue to cause loss and damage that will make many places uninhabitable, and will in some cases make peoples' homelands disappear altogether (e.g. small island nations) (see Christensen 2017, Risse 2009, *Climate Science Special Report* 2017).

There is no question that there is a worsening climate refugee problem. The urgent practical and normative questions to answer are how to determine who should be legally responsible for climate refugees and what are effective and ethical ways of helping such refugees. This chapter examines avenues for realizing effective responses to mounting 'climate refugee' justice issues through UN refugee and climate policy mechanisms. I begin by discussing normative justice issues that are unique to *climate* refugee cases, and explain why climate refugees are not currently counted as refugees – and thus not extended rights to non-refoulement, asylum and relief available to other sorts of refugees – under the UN Refugee Convention (hereafter UNRC). Subsequently, I assess three proposals for addressing climate refugee issues made in the literature on climate justice, and propose two other approaches that can complement those existing proposals. The examined proposals include: (1) revising the criterion for 'refugeedom' and associated provisions of the UNRC to extend to climate refugees the rights and protections of the UNRC; (2) devising a system for issuing emergency passports for climate refugees; (3) pre-emptively relocating people living in places that will soon experience loss and damage due to climate change (e.g. those on small island nations); (4) pursuing responses through the 'risk sharing' and 'loss and damage' provisions of the UN Framework Convention on Climate Change; and (5) using new climate science modelling techniques to assist in devising and prioritizing response strategies. I conclude by reflecting on the ways that these five proposals could be incentivized and co-implemented to help realize effective and ethical international responses to climate refugee justice issues.

The special challenge of climate refugees

Cases of *climate* refugeedom require new and innovative responses because they differ substantively from other sorts of refugeedom. Accordingly, observing distinctions between 'sociopolitical', 'environmental' and 'climate' refugeedom helps in evaluating different ethical, legal and practical obstacles to helping different sorts of refugees. 'Sociopolitical refugees' are those who are forced to flee their homeland and are unable to return because of sociopolitical factors like war or genocide. 'Environmental refugees' are those forced to flee their homeland and who are unable to return due to environmental factors (Lister 2014). 'Climate refugees' are those who are forced to flee their homeland and are unable to return because of events, environmental or non-environmental, arising from climate change (Donhauser 2018). The case of the Peruvian refugees is an example of *environmental* refugees, since floods and crop and water contamination forced them to flee Peru without being able to return and live in humane conditions. Such refugees are also *climate* refugees because the relevant flooding and contamination issues were caused by rapid glacial melt due to historically unprecedented global warming spikes. Thus, 'climate refugee' and 'environmental refugee' class designations overlap. However, they do not overlap completely.

This is because people can become environmental refugees due to events that were not caused by climate change (e.g. industrial or nuclear contamination). People can also become climate refugees for non-environmental reasons. People forced to migrate because of severe drought due to climate change may inhabit a nearby region and thus displace the original inhabitants of that place for example. This has occurred in numerous instances in East Africa (see, for example, van Baalen and Mobjörk 2017). In such cases, refugeedom is caused by political instability and security issues due to events linked to climate change, and would thus be at once cases of climate refugeedom and sociopolitical refugeedom. In other cases, sociopolitical, environmental and climate factors are mutually reinforcing. For instance, we have seen Syrian refugees fleeing because of losses from exceptionally long periods of drought associated with climate change, but who are also fleeing because of safety issues due to political instability and war that are arguably exacerbated by those poor environmental conditions (see Culbertson 2016, Wendle 2016).

Although it is difficult, and in some ways futile, to try to neatly sort cases of refugee-dom into types, it is nevertheless crucially important to make distinctions between different kinds and causes of refugeedom because institutions and legal provisions designed to help refugees must have criteria for whom they are designed to help and how. This becomes clearer upon consideration of the following normative justice issues regarding climate refugees.

Numerous considerations of normative justice hinge on establishing culpability and liability for loss or damage suffered by climate refugees. Unlike cases of socio-political and non-climate-related environmental refugeedom, most cases of climate refugeedom are the result of the impacts of aggregated contributions to greenhouse gas emissions on global warming. And notably, the developed nations that played the most statistically significant causal roles in driving climate change in recent decades (e.g. the US and Western European nations) are also those whose develop-ment has given them the most resources to help climate refugees. Below I explain how new climate science methods permit researchers to assign statistical degrees of liability to the governments of countries for loss and damage due to events linked to climate change. Yet, even a cursory consideration of the situation arguably shows that such nations are liable for knowingly putting people in at least some places at serious risk of becoming climate refugees. Consider, for example, loss and damage resulting from the actions of nations that took those actions when there was ample evidence that they could significantly raise the probability of exactly those kinds of loss and damage.

Take the case of Peruvian climate refugees as an example. Major emitting nations continued to fail to reduce emissions even when scientific consensus showed that doing so would likely cause loss and damage from glacial melting (Christidis et al. 2019, Herring et al. 2019, Otto et al. 2018, cf. Oreskes 2004, Friedrich et al. 2017, Shockley and Boran 2015). For reasons discussed below, many argue that climate refugee cases like this are clear-cut climate justice cases where statistical degrees of culpability can be established. Accordingly, some contend that the major emitting nations that are members of the UN, and subject to offering asylum and protections to refugees under UN laws, are obligated to offer asylum and aid in at least these sorts of cases (see, for example, Huggel et al. 2016, James et al. 2019, Otto et al. 2017, Thompson and Otto 2015). These considerations lead us to another set of normative issues that concern whether climate refugees should be assisted under the legal provisions developed to address refugees *in general*.

Some legal scholars argue that fleeing from *persecution* is a fundamental aspect of being a refugee, which implies that climate refugees are not refugees and should not be extended the protections designed to help refugees (see Shacknove 1985, Price 2009, Cherem 2016). Defenders of this position seek to uphold the current criteria for refugee status given in the UNRC. Those criteria are stipulated in the 1967 Protocol to the UN Convention on the Status of Refugees, which says that refugees are persons who:

[O]wing to a well-founded fear of being persecuted for reasons of race, religion, nationality, membership in a particular social group, or political opinion, is outside the country of his [sic] nationality and is unable or, owing to such fear, is unwilling to avail himself [sic] of the protection of that country; or who, not having a nationality and being outside the country of his [sic] former habitual residence, is unable or, owing to such fear, is unwilling to return to it. (Convention 1967: Art. 1A, 2)

These criteria are maintained in the current UN policy framework (see UNHRC 2017). Hence, it is argued that climate refugees should not be helped through UN channels for assisting refugees because the UN's narrow legal definition of 'refugee' does not count them as refugees.

Some contend that the UN definition should be broadened since there is no reason in principle to link persecution and refugee status. Lister (2016, 2014, 2013), for example, argues that the UN criteria should be broadened to include refugees not counted under the 1967 protocol by de-emphasizing persecution and asylum within the definition. He proposes that refugee status is warranted whenever there is dislocation due to, 'disruption of expected indefinite duration, where international movement is necessitated, and where the threat is not just to a favoured or traditional way of life, but to the possibility of a decent life at all' (2014: 4, 2016: 51–2). Lister's criteria would recognize what I have distinguished as climate, environmental and sociopolitical refugees as refugees under the UNRC, such that they would have rights to asylum and non-refoulement granted under the Convention.

Lister has argued, moreover, that the need for asylum need not be emphasized as there are other ways to help refugees besides providing asylum through the provisions of the UNRC, like providing relief aid and military support. In response, Cherem (2016) has contended that the UN definition of 'refugee' should not be expanded to include climate refugees because that would be impractical. He argues that the legal provisions and unilateral protection measures that currently pertain to refugees under the UNRC would be less effective than alternative ways of seeking justice for climate refugees. Of course, this proposal begs the main question this chapter seeks to answer: What are the potential ways of effectively helping climate refugees to realize climate justice?

Proposals for helping climate refugees

Changing the legal definition of 'refugee'

Lister's suggestion to revise the criteria for refugeedom in the UNRC is one simple way to begin providing rights to asylum, non-refoulement and relief options to climate refugees. Since there is insufficient reason to uphold the current criteria, I submit that this can and should happen for practical reasons discussed below. Still, Lister's proposal has notable drawbacks. For starters, the provisions of the UNRC will be insufficient on their own to enable adequate responses to climate refugee justice issues because of the inevitable increasing magnitude and severity of climate refugee cases. The UNRC does not contain any protocol for dealing with such severe problems, because its criteria are an artefact of responding to relatively rare and clear-cut cases where sociopolitical refugees sought asylum after fleeing war and persecution. Another drawback is that revising the criteria will require international negotiations between all UN parties, which would be difficult and generally takes decades. Yet there are already climate refugees, and it is likely that there will be many more in the near future. It is thus crucially important to devise complementary responses to climate refugee cases more immediately (Donhauser 2018: 11).

Emergency passports

A potential way to help climate refugees that does not require retooling the UNRC is to issue climate refugees emergency 'Nansen' passports. Nansen passports were the first issued to peoples who became stateless due to forced migrations or unclear international laws, to allow them to lawfully enter and seek asylum in relatively safer countries. Clare Heyward and Jörgen Ödalen (2016) have argued that climate refugees meet the criteria for those who have been issued Nansen passports in the past, and propose that climate refugees be issued a new sort of Nansen passport they call a 'Passport for the Territorially Disposed'. Since this would be a quicker way to grant climate refugees rights to asylum and temporary non-refoulement than modifying the UNRC, I submit that it can and should be pursued as a complementary response. However, this proposal faces practical problems that render it insufficient on its own.

Beyond having to pass through UN negotiations, individual countries will need to devise a protocol and allocate funding for handling individuals issued such passports. Furthermore, like the proposal to revise the UNRC, clear threshold criteria for issuing such passports will need to be devised (Donhauser 2018: 12). The passports would also provide a very limited solution if implemented, since passports do not on their own permit people to stay in foreign countries for extended periods of time. So, while this proposal should be pursued as well, it is nonetheless crucial to seek revision of the UNRC and further complementary responses to help realize justice for climate refugees.

Pre-emptive relocation

Anote Tong, President of the island of Kiribati, proposed a further complementary response at the 2008 UN General Assembly. As one of several small island nations, Kiribati's residents have been watching the ocean slowly swallow their homeland due to climate-driven sea-level rise, and already suffer losses (e.g. drastically increased infant mortality) due to potable water salinization and land loss. Tong asked the UN Assembly to consider offering pre-emptive relief through a systematic relocation of his people so that, in his words, 'when [my] people migrate, they will migrate on merit and with dignity' (United Nations 2008). Risse (2009) considers ways of justifying and motivating such a pre-emptive strategy for addressing inevitable climate refugeedom. He proposes not just relocation, but also systematic large-scale redistribution of people in imminent danger of becoming climate refugees throughout the world. Here the idea is to incentivize such a pre-emptive response by fairly distributing climate refugees so that no single nation has to shoulder an unfair share of the burden of taking in such refugees. Risse's proposal broaches the general implementation details of a way to seek justice for climate refugees and, in Kiribati, has begun to be realized. Through a multi-phase relocation and adaption plan, in cooperation with the World Bank and other partners, residents who wish to relocate have been enabled to do so – primarily to Australia, Banaba Island, Nauru and New Zealand (Kiribati Adaptation Program n.d., Oakes et al. 2016: 26). One problem of such a response, which is visible in the case of Kiribati, is that many people will choose to remain in their homeland, and will therefore continue to suffer unjust loss and damage (see Walker 2017).

Yet, even if these injustices can be offset with additional forms of relief, certain practical considerations must be addressed to implement such responses for other climate refugee cases. Two considerations are especially important. First, realizing such pre-emptive responses requires determining what institutional and economic channels can be used to do so in particular cases. Second, systematic criteria and means of evaluating risks of populations being more or less likely to experience imminent loss or damage are needed to determine and prioritize where to enact such responses (Donhauser 2017: 266–7, 2018: 13–14).

'Risk sharing' and the UN Framework Convention on Climate Change

A proposal that is responsive to the aforementioned concerns about finding institutional and economic channels for implementing responses to emerging climate refugee cases is to utilize certain provisions of the 1992 United Nations Framework Convention on Climate Change (hereafter UNFCCC) and the 2015 Paris Agreement on climate change (UNFCCC 2015). The UNFCCC and the policy track leading to it focused explicitly on taking steps to implement adaptive responses at all levels to address emerging and imminent problems of climate change. By the end of 2017, 168 parties had ratified the agreement and officially began taking steps to collectively reduce global current net emissions and enact 'climate action plans' (Paris Agreement Ratification Tracker 2017). Though, at present, it is too abstract and not yet fleshed out enough to enable specific solutions to particular climate justice issues, the UNFCCC nevertheless provides a foundation for the legal and political infrastructure necessary for realizing ethical and effective international responses to climate justice issues.

Of particular interest here are the 'risk sharing' provisions contained in the 'Losses and Damages' section of the UNFCCC. That section directs specified committees to develop international 'risk assessment and management [sic] risk insurance facilities, climate risk pooling and other insurance solutions' to respond to loss and damage due to climate change (UNFCCC 2015: Article 8; UNFCCC 2013; see COP24 (2018) for progress). This risk-insurance pool is supposed to be a source of funds pooled by parties to the agreement that can be allocated to address loss or damage suffered by any party to the agreement going forward. Once operational, this pool could thus be used to realize and fund efforts such as the emergency passport and pre-emptive relocation responses discussed above (Donhauser 2017: 272–4, 2018: 14). The insurance pool could also be used to mitigate loss and damage experienced by climate refugees through distribution of relief and aid funds.

A major advantage of pursuing responses through these provisions of the UNFCCC is that the agreement is already in effect, and, though some parties are not fully in compliance with the agreement to date, all but one of leading governments in the world have begun taking steps to comply with their commitments under the agreement. Pursuing responses through the UNFCCC thus provides a way to side-step some of the practical hurdles to pursuing responses through the UN Refugee Convention. Still, making the UNFCCC risk-insurance operational, and figuring out how to use it to address particular climate justice issues ethically and effec-tively, will require a lot of work going forward. Finding means of devising criteria and risk thresholds for prioritizing where to distribute relief and where it would be best to pursue pre-emptive responses will be crucially important to finding ways to effectively implement the provisions of the UNFCCC. My final proposal responds to this need for such criteria and risk thresholds.

Using Probabilistic Event Attribution Studies to help realize climate justice

Numerous climate scientists and ethicists have argued that the methods of Probabilistic Event Attribution studies (PEAs) are useful for decision-making about climate justice issues (see Allen 2003, Donhauser 2017, Otto et al. 2017, Thompson and Otto 2015). These studies use 'super-ensemble' atmospheric data models and statistical analyses to identify major causes of extreme weather events. Super-ensemble weather prediction models are a set of very many simulations, a disjunctive set, generated from the same data to assess what is the most statistically probable range of predictions.

PEA researchers generate these models by varying the variables and parameters of independent models used to run simulations until they accurately simulate an extreme weather event that actually occurred. They then systematically remove select variables (like US or French emissions during a certain range of years) until the weather event does not occur in simulations with that same most accurate model. Through comparisons of accurate simulations of an event and simulations in which it doesn't occur when various climate drivers are removed, they then calculate the statistical probability of the salient extreme weather event occurring when certain climate change drivers are absent (see, for example, Bindoff et al. 2013, James et al. 2014, Stott et al. 2004, Stott et al. 2013, Stone and Allen 2005, Stone et al. 2013). Researchers can provide objective justification for claims that certain events could not have happened if certain major climate change drivers had not been present with such studies. For example, they could be used to show that droughts in Syria or floods in Peru or Puerto Rico would not have occurred had US emissions, and greenhouse warming potential, been statistically lower during the years directly preceding these events (cf. Donhauser 2017: 263–4, Otto et al. 2017).

To be clear, PEAs do not provide grounds for assigning sole liability for any particular event linked to climate change to any one country, because global climate change is caused through aggregated causes (cf. Biber 2008). Still, these studies can be, and are, used to establish which nations played statistically larger or smaller roles in bringing about particular events. They can thus be used to establish percentages of proportional liability for loss and damage due to events linked to climate change, and they can help to 'slice up the liability pie' formed via the UNFCCC's risk-pooling provisions. Arguably, PEAs can also be used to help prorate risk-insurance contributions and distributions of funds from the risk-insurance pool for climate caused loss and damage that has already occurred (Thompson and Otto 2015). In these ways, PEAs could be used to help realize just responses to loss and damage experienced by climate refugees (Donhauser 2018: 13). Thompson and Otto (2015) and Otto et al. (2017) present arguments along these lines. They argue that since 'historical responsibility of individual countries and regions can now be quantified for specific extreme events' with PEAs, such studies can be used to assign 'statistical liability' to individual countries for purposes of litigation to address climate justice issues that have already occurred (Otto et al. 2017: 759, cf. Bhattacharya 2003, Field and Barros 2014, Hulme 2014, Peterson et al. 2012, Spross 2014).

What is more, it is arguable that PEAs can be used to help make estimations about the impacts of potential future extreme weather events. Because such studies are 'enhancing knowledge about changes in extreme weather event patterns, intensity and frequency under different climate regimes', they reveal information about how weather events and their causes that can be used to help make decisions about pre-emptive and mitigatory efforts to minimize loss and damage (Donhauser 2017: 265, cf. Funk 2012, Luo et al. 2015, Yu et al. 2014). Simulations used in PEAs can also be used to help estimate likelihoods of potential future extreme events and their possible impacts under a range of probable future climate conditions (Donhauser 2017: 265, cf. Dole et al. 2011, Otto et al. 2012, Pall et al. 2011, Peterson and Heim 2013, Rahmstorf and Coumou 2011, Rozenberg et al. 2014, Yzer et al. 2014). Further still, such simulations can also be used to garner insights into how the presence or absence of certain climatic changes affect the likelihoods of weather events that can impact valued resources (e.g. potable water and agriculture) (Donhauser 2017: 266; see also Challinor et al. 2009, Dumas and Ha-Duong 2013, Kurukulasuriya et al. 2011, Seo and Mendelsohn 2008). The simulations used in PEAs could thus be used to help estimate which peoples will have greater or lesser degrees of risk of experiencing loss or damage due to coming extreme weather events (cf. Donhauser 2017: 266–7, 2018: 13). In this way too, they could help to prioritize where pre-emptive mitigation and relief efforts would be most reasonably allocated.

Conclusion

There are many avenues for additional research to help realize justice for climate refugees. For starters, exploring ways to effectively operationalize the discussed proposals for getting asylum for climate refugees – through revising the UNRC, by using the provisions of the UNFCCC, by issuing special Nansen passports, or all three – is an avenue for work that will be crucially useful to policymakers. There is also valuable work to be done on ways that those proposals could be incentivized. For example, it would be useful to examine whether offering asylum to climate refugees could be incentivized by treating such actions as contributions to the UNFCCC risk-insurance pool that would offset monetary contributions. It would also be useful to assess potential ways of fairly distributing climate refugees, and people in imminent danger of experiencing loss and damage, so that certain nations are not taking on more burden than others. For instance, it is important to explore whether distributing such refugees should reflect each participating country's proportional liability for contributing to the risk-insurance pool.

Exploring more specific applications of PEAs for evaluating statistical liability, possible future scenarios and risk potentials, is another avenue for research that could be vital to realizing effective and ethical responses to climate justice issues. Accordingly, there are many questions, and live debate, about how PEAs can be used to operationalize the risk-insurance and pooling provisions of the UNFCCC Paris Agreement. There are also related questions about whether and how they could be used to help realize forms of justice for climate refugees (cf. Donhauser 2017, Thompson and Otto 2015).

Another important area for further research concerns worries about the right of climate refugees to choose where they are able to be relocated. Providing those experiencing climate refugeedom options, and the autonomy to choose whether and where they will be relocated, is arguably essential to realizing ethically acceptable responses. For instance, if it were determined that countries with the highest emissions were obligated to offer asylum to the greatest number of climate refugees, such refugees may be better off, according to their own standards, emigrating to certain low-emitting countries like Nauru, New Zealand, Sweden or the Netherlands.

Finally, there may be concerns regarding my suggestion to pursue pluralistic solutions to climate justice issues, and specifically regarding my proposals as complementary ways of realizing justice for climate refugees. A common worry, for instance, is that pursuing any pluralistic set of solutions will entail putting less effort and resources into developing any particular one of those solutions effectively. The corpus of existing work in ethics, decision theory and related areas provides rich resources for beginning to address such concerns, as well as those discussed above. There are no easy solutions to any of these concerns. Realizing effective and ethical responses to cases of climate refugeedom and other forms of climate injustice will require further critical discussion, debate and planning. It is my sincere hope that this chapter will help to facilitate those responses.

Acknowledgements

Thanks to Gillian Barker, Max Cherem, Eric Desjardins, Allen Thompson, John Corcoran, Jamie Shaw and students in my 2018 Environmental Ethics course at Bowling Green State University for comments and helpful discussions during the development of parts of this chapter.

References

Allen, Myles (2003), 'Liability for climate change', *Nature*, 421(6926), 891–2.

Bhattacharya, Shaoni (2003), 'European heatwave caused 35,000 deaths', *New Scientist*, 10(10), 03.

Biber, Eric (2008), 'Climate change, causation and delayed harm', *Hofstra Law Review*, 37, 975.

Bindoff, Nathaniel L., Peter Stott, Krishna Mirle, Achuta Rao, Myles R. Allen, Nathan Gillett, David Gutzler et al. (2013), 'Detection and attribution of climate change: from global to regional', in T.F. Stocker, D. Qin, G-K. Plattner, M. Tignor, S.K. Allen, J. Boschung, A. Nauels et al. (eds), *Climate Change 2013: The Physical Science Basis, Contribution of Working Group I to the Fifth Assessment Report of the Intergovernmental Panel on Climate Change*, Cambridge: Cambridge University Press, pp. 867–928.

Challinor, Andrew Juan, Tim Wheeler, Debbie Hemming and H.D. Upadhyaya (2009), 'Ensemble yield simulations: crop and climate uncertainties, sensitivity to temperature and genotypic adaptation to climate change', *Climate Research*, 38(2), 117–27.

Cherem, Max (2016), 'Refugee rights: against expanding the definition of a "refugee" and Unilateral Protection Elsewhere', *Journal of Political Philosophy*, 24(2), 183–205.

Christensen, Jen (2017), '16,000 scientists sign dire warning to humanity over health of planet', *CNN*, accessed 17 October 2018 at: https://www.cnn.com/2017/11/14/health/scientists-warn-humanity/index.html.

Christidis, N., Richard Betts and Peter Stott (2019), 'The extremely wet March of 2017 in Peru', *Bulletin of the American Meteorological Society*, 100(1), S31–S35.

Climate Change and Disasters (2015), *United Nations High Commissioner for Refugees*, accessed 29 October 2018 at: http://www.unhcr.org/en-us/climate-change-and-disasters.html.

Climate: observations, projections and impacts (2011), UK Met Office, accessed 15 June 2017 at: https://www.metoffice.gov.uk/climate-guide/science/uk/obs-projections-impacts.

Climate Refugee, *National Geographic*, accessed 11 July 2017 at: https://www.nationalgeographic.org/encyclopedia/climate-refugee/.

Climate Science Special Report: Fourth National Climate Assessment (2017), USGCRP, accessed 5 July 2018 at: https://science2017.globalchange.gov/.

Congress of Parties (COP) 24 (2018), *United Nations: Climate Change*, accessed 1 November 2019 at: https://unfccc.int/event/cop-24.

Convention and Protocol Relating to the Status of Refugees (1967), accessed 1 November 2019 at: https://www.unhcr.org/3b66c2aa10.html.

Culbertson, Shelly (2016), 'A different kind of refugee crisis', *U.S. News and World Report*, accessed 17 May 2016 at: https://www.usnews.com/opinion/articles/2016-05-16/aid-community-must-rethink-approach-to-syrian-refugees-in-the-middle-east.

Dole, Randall, Martin Hoerling, Judith Perlwitz, Jon Eischeid, Philip Pegion, Tao Zhang, Xiao-Wei Quan, Taiyi Xu and Donald Murray (2011), 'Was there a basis for anticipating the 2010 Russian heat wave?', *Geophysical Research Letters*, 38(6), 1–5.

Donhauser, Justin (2017), 'The value of weather event science for pending climate policy decisions', *Ethics, Policy and Environment*, 20(3), 263–78.

Donhauser, Justin (2018), 'How new climate science and policy can help climate refugees', *Journal of Ethical Urban Living*, **1**(2), 1–21.

Dumas, Patrice and Minh Ha-Duong (2013), 'Optimal growth with adaptation to climate change', *Climatic Change*, **117**, 691–710.

Field, Christopher and Vicente Barros (eds) (2014), *Climate Change 2014: Impacts, Adaptation and Vulnerability. Part A: Global and Sectoral Aspects, Contribution of Working Group II to the Fifth Assessment Report of the Intergovernmental Panel on Climate Change*, Volume 1, Cambridge: Cambridge University Press.

Friedrich, Johannes, Mengpin Ge and Andrew Pickens (2017), 'This interactive chart explains world's top 10 emitters and how they've changed', *World Resources Institute*, accessed 25 December 2018 at: http://www.wri.org/blog/2017/04/interactive-chart-explains-worlds-top-10-emitters-and-how-they ve-changed.

Funk, Chris (2012), 'Exceptional warming in the Western Pacific-Indian Ocean warm pool has contributed to more frequent droughts in eastern Africa', *Explaining extreme events of 2011 from a climate perspective: Bulletin of the American Meteorological Society*, **93**, 1049–51.

Herring, Stephanie, Nikolaos Christidis, Andrew Hoell, Martin P. Hoerling and Peter A. Stott (eds) (2019), 'Explaining extreme events of 2017 from a climate perspective', Special Supplement to the *Bulletin of the American Meteorological Society*, **100**(1).

Heyward, Clare and Jörgen Ödalen (2016), 'A free movement passport for the territorially dispossessed', *Climate Justice in a Non-Ideal World*, Oxford Index, 208–226, accessed 3 April 2019 at: http://oxfordin dex.oup.com/view/10.1093/acprof:oso/9780198744047.003.0011.

Huggel, Christian, Ivo Wallimann-Helmer, Dáithí Stone and Wolfgang Cramer (2016), 'Reconciling justice and attribution research to advance climate policy', *Nature Climate Change*, **6**(10), 901–908.

Hulme, Mike (2014), 'Attributing weather extremes to "climate change": a review', *Progress in Physical Geography*, **38**(4), 499–511.

James, Rachel, Friederike Otto, Hannah Parker, Emily Boyd, Rosalind Cornforth, Daniel Mitchell and Myles Allen (2014), 'Characterizing loss and damage from climate change', *Nature Climate Change*, **4**(11), 938–9.

James, Rachel A., Richard G. Jones, Emily Boyd, Hannah R. Young, Friederike E.L. Otto, Christian Huggel and Jan S. Fuglestvedt (2019), 'Attribution: how is it relevant for loss and damage policy and practice?', in R. Mechler, L.M. Bouwer, T. Schinko, S. Surminski and J. Linnerooth-Bayer (eds), *Loss and Damage from Climate Change: Concepts, Methods and Policy Options*, Cham: Springer International Publishing, pp. 113–54.

Kiribati Adaptation Program (n.d.), *Kiribati: Climate Change*, accessed 1 January 2019 at: http://www. climate.gov.ki/category/action/relocation/.

Kurukulasuriya, Pradeep, Namrata Kala and Robert Mendelsohn (2011), 'Adaptation and climate change impacts: a structural Ricardian model of irrigation and farm income in Africa', *Climate Change Economics*, **2**(02), 149–74.

Lister, Matthew (2013), 'Who are refugees?', *Law and Philosophy*, **32**(5), 645–71.

Lister, Matthew (2014), 'Climate change refugees', *Critical Review of International Social and Political Philosophy*, **17**(5), 618–34.

Lister, Matthew (2016), 'The place of persecution and non-state action in refugee protection', in A. Sager (ed.), *The Ethics and Politics of Immigration: Core Issues and Emerging*, Lanham, MD: Rowman and Littlefield, pp. 45–60.

Luo, Pingping, Bin He, Kaoru Takara, Yin E. Xiong, Daniel Nover, Weili Duan and Kensuke Fukushi (2015), 'Historical assessment of Chinese and Japanese flood management policies and implications for managing future floods', *Environmental Science and Policy*, **48**, 265–77.

Oakes, Robert, Andrea Milan and Jillian Campbell (2016), 'Kiribati: climate change and migration – relationships between household vulnerability, human mobility and climate change', *Bonn: United*

Nations University Institute for Environment and Human Security, Report No. 20, accessed 4 April 2019 at: http://collections.unu.edu/eserv/UNU:5903/Online_No_20_Kiribati_Report_161207.pdf.

Oreskes, Naomi (2004), 'The scientific consensus on climate change', *Science*, **306**, 1686.

Otto, Friederike, N. Massey, G.J. van Oldenborgh, R.G. Jones and Myles Allen (2012), 'Reconciling two approaches to attribution of the 2010 Russian heat wave', *Geophysical Research Letters*, **39**(4), 1–5.

Otto, Friederike, Ragnhild B. Skeie, Jan S. Fuglestvedt, Terje Berntsen and Myles Allen (2017), 'Assigning historic responsibility for extreme weather events', *Nature Climate Change*, **7**, 757–9.

Otto, Friederike, Sjoukje Philip, Sarah Kew, Sihan Li, Andrew King and Heidi Cullen (2018), 'Attributing high-impact extreme events across timescales – a case study of four different types of events', *Climatic Change*, **149**(3–4), 399–412.

Pall, Pardeep, Tolu Aina, Dáithí A. Stone, Peter A. Stott, Toru Nozawa, Arno G.J. Hilberts, Dag Lohmann and Myles R. Allen (2011), 'Anthropogenic greenhouse gas contribution to flood risk in England and Wales in autumn 2000', *Nature*, **470**(7334), 382–5.

Paris Agreement Ratification Tracker (2017), accessed 17 October 2018 at: http://climateanalytics.org/briefings/ratification-tracker.html.

Peterson, Thomas C. and Richard R. Heim Jr (2013), 'Monitoring and understanding changes in heat waves, cold waves, floods and droughts in the United States: state of knowledge', *Bulletin of the American Meteorological Society*, **94**(6), 821–34.

Peterson, Thomas C., Peter A. Stott and Stephanie Herring (2012), 'Explaining extreme events of 2011 from a climate perspective', *Bulletin of the American Meteorological Society*, **93**(7), 1041–67.

Price, M. (2009), *Rethinking Asylum: History, Purpose, and Limits*, Cambridge: Cambridge University Press.

Rahmstorf, Stefan and Dim Coumou (2011), 'Increase of extreme events in a warming world', *Proceedings of the National Academy of Sciences*, **108**(44), 17905–909.

Risse, Mathias (2009), 'The right to relocation: disappearing island nations and common ownership of the earth', *Ethics and International Affairs*, **23**(3), 281–300.

Rozenberg, Julie, Céline Guivarch, Robert Lempert and Stéphane Hallegatte (2014), 'Building SSPs for climate policy analysis: a scenario elicitation methodology to map the space of possible future challenges to mitigation and adaptation', *Climatic Change*, **122**(3), 509–22.

Seo, S. Niggol and Robert Mendelsohn (2008), 'An analysis of crop choice: adapting to climate change in South American farms', *Ecological Economics*, **67**(1), 109–16.

Shacknove, Andrew E. (1985), 'Who is a refugee?', *Ethics*, **95**(2), 274–84.

Shockley, Kenneth and Idil Boran (2015), 'COP 20 Lima: the ethical dimension of climate negotiations on the way to Paris – issues, challenges, prospects', *Ethics, Policy and Environment*, **18**(2), 117–22.

Spross, Jeff (2014), 'Heat waves could triple premature deaths in Britain by 2050', accessed 14 May 2016 at: http://thinkprogress.org/climate/2014/07/09/3458126/britain-heat-hospitals/.

Stone, Dáithí A. and Myles Allen (2005), 'The end-to-end attribution problem: from emissions to impacts', *Climatic Change*, **71**(3), 303–18.

Stone, Dáithí A., Maximilian Auffhammer, Mark Carey, Gerrit Hansen, Christian Huggel, Wolfgang Cramer, David Lobell et al. (2013), 'The challenge to detect and attribute effects of climate change on human and natural systems', *Climatic Change*, **121**(2), 381–95.

Stott, Peter A., Dáithí Stone and Myles Allen (2004), 'Human contribution to the European heatwave of 2003', *Nature*, **432**(7017), 610–14.

Stott, Peter A., Myles Allen, Nikolaos Christidis, Randall M. Dole, Martin Hoerling, Chris Huntingford, Pardeep Pall et al. (2013), 'Attribution of weather and climate-related events', in G.R. Asrar and J.W. Hurrell (eds), *Climate Science for Serving Society*, New York: Springer, pp. 307–37.

Thompson, Allen and Friederike Otto (2015), 'Ethical and normative implications of weather event attribution for policy discussions concerning loss and damage', *Climatic Change*, **133**(3), 439–51.

UNFCCC (2013), 'Warsaw international mechanism for loss and damage associated with climate change impacts', UNFCCC, accessed 25 December 2017 at: http://unfccc.int/resource/docs/2013/cop19/eng/10a01.pdf.

UNFCCC (2015), 'Adoption of the Paris Agreement', UNFCCC, accessed 25 December 2017 at: http://unfccc.int/resource/docs/2015/cop21/eng/l09r01.pdf.

UNHRC (2017), 'What is a refugee?', *Refugees Facts*, accessed 16 April 2019 at: https://www.unrefugees.org/refugee-facts/what-is-a-refugee/.

United Nations (2008), 'Reeling from impacts of global warming, small island states urge General Assembly to take comprehensive action', United Nations, accessed 10 July 2011 at: http://www.un.org/press/en/2008/ga10754.doc.htm.

van Baalen, Sebastian and Malin Mobjörk (2017), 'Climate change and violent conflict in East Africa', *International Studies Review*, 20(4), 547–75.

Walker, Ben (2017), 'An island nation turns away from climate migration, despite rising seas', *Inside Climate News*, accessed 11 January 2019 at: https://insideclimatenews.org/news/20112017/kiribati-climate-change-refugees-migration-pacific-islands-sea-level-rise-coconuts-tourism.

Wendle, John (2016), 'The ominous story of Syria's climate refugees', *Scientific American*, March, accessed 20 May 2016 at: https://www.scientificamerican.com/article/ominous-story-of-syria-climate-refugees/.

Yu, Meixiu, Qiongfang Li, Michael Hayes, Mark Svoboda and Richard Heim (2014), 'Are droughts becoming more frequent or severe in China based on the standardized precipitation evapotranspiration index: 1951–2010?', *International Journal of Climatology*, 34(3), 545–58.

Yzer, Jerrel R., Warren E. Walker, Vincent A.W.J. Marchau and Jan H. Kwakkel (2014), 'Dynamic adaptive policies: a way to improve the cost–benefit performance of megaprojects', *Environment and Planning B: Planning and Design*, 41(4), 594–612.

11 Pre-emptive justice for future generations: reframing climate change as a 'humanitarian climate crime'

Selina Rose O'Doherty

Climate change is an accumulative and transnational phenomenon, in which the causes occur distantly across time and space from the effect itself. Based on these properties, it is widely considered that the victims of the most severe effects of climate change will predominantly be those who have done the least to create it (Page 2007, Samson et al. 2011). Therefore, future persons (both inter- and intra-generational persons) who are separated from the causes of climate change by time will disproportionately suffer the effects of it via their inherited environment. As such, it seems fair to claim that future generations are suffering a climate injustice, given that the actions which will create ongoing threats to their future human security are being knowingly carried out by current generations. It is worth noting that the effects considered within the scope of this argument are those which are considered, according to current knowledge, to be definite outcomes of climate change if current global practices continue as-is, rather than merely possible or probable outcomes. In short, according to the accepted research, the consequences of climate change discussed throughout this chapter are those which are guaranteed to happen. Although harming someone directly is a crime, and harming subjects en masse is considered a public harm, harming subjects at a future point in time, even if the harm is caused by those same current and deliberate actions, is not pre-emptively considered a crime. This is despite awareness of the consequences of current actions passing the irreversibility threshold and rendering catastrophic climate change and its subsequent harms unavoidable – for example, the melting of the West Antarctic ice sheet cannot be prevented (McCusker et al. 2015).

In this chapter I assume that current generations hold certain climate justice obligations to future generations and I posit that, due to the characteristics of climate change, meeting this obligation requires taking pre-emptive action. The case also builds on the well-established premises that: people must have certain basic physical needs met in order to live, which require viable air, soil and water; the planet and the human race will still exist, in recognizable form, by the end of this century; and the global political order takes the form of an international society as defined by the English School of International Relations theory (Linklater 2005, Evans and Newnham 1998: 148). Additionally, the responsible actors referred to throughout this chapter are state and intergovernmental actors, which are deemed to possess an overarching legitimate authority over the numerous other actors carrying out the causal actions of climate change daily on a micro-level (for a detailed defence of this position on state responsibility, see O'Doherty 2018). The approval by all 193 United Nations (UN) member states (and both UN observers) of the published reports of the Intergovernmental Panel on Climate Change (IPCC) is taken here to demonstrate acceptance – and thus to indicate an awareness by all approving states – of anthropogenic climate change and the predicted consequences. Despite the widening gap between the science and the politics of climate change which sees climate change denial from some leading political actors – such as Brazilian President Jair Bolsonaro or United States President Donald Trump – the original approval for the recent IPCC (2018) report implies that the respective authorities are aware of the content of the reports.

Given that good intentions and enthusiastic planning cannot, at the current rate and in the current political climate, prevent forthcoming catastrophic environmental injustices, I suggest a reinterpretation of existing practices which may be applied to compel states to take the actions necessary to tackle climate change. Using, among other instruments, international human rights law and the International Criminal Court (ICC) to prevent the causal behaviours of climate change offers an alternative way to prevent catastrophic climate harms from occurring. Rather than necessitating a new set of specific environmental rights or creating new laws, legal frameworks or systemic norms, for which there is arguably not enough time, the necessary pre-emptive action can be justified and employed through a reframing of the causal actions of climate change as humanitarian climate crimes – of being deliberate, and thus criminal, harms to those who will suffer the clear and present danger of the predicted effects of climate change. In short, the idea of humanitarian climate crimes implies accountability for causing a humanitarian disaster via the causal practices of climate change.

The significant departure that my argument takes from the current application of existing mechanisms to protect against human rights violations and crimes against humanity is in the element of pre-emption. Generally speaking, an actor cannot be found guilty of a crime against humanity or a gross human rights violation in advance of the consequences of the causal action occurring, However, as climate change has the unusual property of being irreversibly set in motion – thus guaranteed to happen – in advance of its actual occurrence, and the causes are identifiable in advance of the irreversibility thresholds being passed, it is essential to pre-emptively consider the causes as being akin to the consequential crimes in order to prevent the injustice of climate change catastrophes. To reframe the causes of climate change in this way, rather than simply acknowledging them as harmful behaviour, and to justify invoking pre-emptive action, two key points must be proven: first, that the forthcoming consequences, if deliberate, would qualify as recognized crimes, and, second, that lack of action to prevent climate change, despite the awareness that state actors have of these consequences, is parallel to a deliberate intent to cause them. Here I will briefly outline the examples of how future climate change-based disaster, if recognized as deliberate, albeit indirectly human-made, are large-scale human rights violations, and thus crimes against humanity, before justifying the framing of awareness as intent.

Humanitarian climate crimes

If caused deliberately, the effects of climate change would be considered breaches of global justice. However, its indirect and delayed nature sees the causal actions being identified but not treated as deliberate harm doing. For example, climate change is predicted to cause ocean levels to rise and completely submerge some low-lying island states (Storey and Hunter 2010). Kiribati, a Pacific island state comprised of 32 small islands and lying almost entirely at sea level, faces submersion of its entire state territory if current causal behaviour continues. If this happens, its permanent population of roughly 110 000 people would be permanently displaced, with threats to the lives, culture and heritage of both the current and successor generations of the i-Kiribati (Vaha 2015). If Kiribati were to be deliberately submerged for the explicit purpose of displacing the population and carrying out a cultural genocide of its people, doing so would be considered criminal under both international human rights law and a crime against humanity. It is these causal actions of climate change, and the indirect intent of the state actors taking these actions, which qualify as either breaches against international human rights law or are definable as crimes against humanity. Thus, those actions can be framed as humanitarian climate crimes.

Violations of human rights

Human rights by their very nature are universal. This is not to say that every right automatically applies to every individual equally, but that every subject to whom a particular right applies is entitled to that right by virtue of their being human, regardless of gender, religion, race, nationality, socio-economic status and so on. However, there is no legal or political status afforded independently to *environmental* rights, despite a sizeable collection of literature arguing for them (Sax 1990, Nickel 1993, Shelton 2006, Francioni 2010, Ramnewash-Oemrawsingh 2011). Creating a framework of humanitarian climate crimes does not necessitate creating such a stand-alone branch of rights, as the interests of future subjects which will be affected by climate change are already protected under current rights and practices. Human rights as recognized in international law are held here to be United Nations sanctioned rights. Although these rights are universal, states are essentially the actors who provide, protect or pursue them for their citizens, through governance. These rights, in various declarations (for example, in addition to the Universal Declaration of Human Rights (United Nations 1948), there is the Fourth Geneva Convention (1949), the European Convention on Human Rights (Council of Europe 1950) and the International Covenant on Social, Political and Cultural Rights (UN 1966)), impose three obligations on states parties: to respect, protect and fulfil the recognized rights (ICISS 2001, WHO 2013) for their own citizens and residents, and in certain circumstances, such as where another state cannot or will not meet its obligations, to intervene on behalf of citizens and residents of these other states.

The UN Human Rights Commission has explicitly stated that climate change impacts the realization of human rights (OHCHR 2009: A/HRC/10/61), indicating that there is an obligation on states to tackle climate change in order to fulfil their human rights obligations. For example, the right to food and shelter, contained within the right to an adequate standard of living (UN 1948: Article 25), is affected by the availability of viable air, soil and water; cultural and language rights are also violated via forced migration or displacement due to climate change; and fulfilling the right to life itself, including the entitlement to protection from clear and present danger, is dependent on provision or pursuit of viable air, soil and water. Therefore, human security (Owens 2012: 547), the provision of which is also an obligation of states to their citizens, is also dependent on physical environment. In the absence of viable air, soil and water, human beings physically cannot thrive; no security can be considered provided and no rights can be met without the presence of these three components (Collins-Chobanian 2000: 135–45). As such, any deliberate threat to their provision is a threat to human security and, by extension, to basic human rights. Threats or harms to soil, in the form of an entire national territory, such as in the case of total submersion caused by rising sea levels, also construes a more traditional threat against sovereign territory, leading to forced migration and violation of the associated human rights. Thus, in order to meet their obligations, all states must pursue or provide protection against climate change for their citizens, including future generations, as well as, where necessary, for citizens of other states.

Other indirect ways in which climate change affects basic subsistence rights are numerous; through malnutrition, pandemics, the spread of disease, water pollution caused by increased flooding, or through an increase in temperature affecting health directly, or via agriculture and the linked threats to nutrition and food security. For example, according to the World Health Organization (WHO), 6.3 per cent of all deaths worldwide in 2008 were attributable to unsafe water (Mahendra et al. 2018), illustrating that the right to water is linked not only to a general human security threat but also to the rights to health and to protection from clear and present danger. Or consider the right to health (UN 1948: Article 25/1) more broadly: it is a fundamental human right whose first articulation predates the universal declaration of human rights. Initially declared in the preamble to the WHO's 1946 constitution as 'a state of complete physical, mental and social well-being and not merely the absence of disease or infirmity', it also states that 'the enjoyment of the highest attainable standard of health is one of the fundamental rights of every human being without distinction of race, religion, political belief, economic or social condition'. It is recognized as a human right in the UN Declaration of Human Rights (UN 1948: Article 25/1) as part of the right to an adequate standard of living and was further entrenched as a human right in the International Covenant on Economic, Social and Cultural Rights (UN General Assembly 1966). It is an inclusive and extensive right, extending into the right or entitlements to safe drinking water, adequate food and basic housing, all of which are adversely harmed by the consequences of climate change. A permanent rise in the average global temperature along with the heatwaves associated with the instability caused by climate change can have an immediate and devastating effect on health, as exemplified during the 2018 global heatwave, which recorded a spike in deaths across Europe, Asia and North America (BBC 2018, Palmer 2018, PHE 2019). Harming a person's health via their environment can be either direct, such as through the release of pollutants directly into that person's immediate environment, or indirect, through harmfully impacting the viability of air, soil or water via climate change, which will in turn cause health crises.

Further to this, of various treaties outlining supranational policy on health in the European Union (EU) (Lazzari et al. 2015), Article 130(R) of the Maastricht Treaty (1992) obliges pre-emptive action, specifying that the objective of the EU environmental policy is to 'protect human health'. Based on the forthcoming climate harms and given that this particular treaty is binding, all EU states have an obligation to refrain from doing or sanctioning any action which will contribute to climate change disasters because doing so is a violation of future persons' human rights to health. Similarly, the right to an adequate standard of living is an inextricable right comprised of, among others, the rights to health, water, food, social security and protection of cultural heritage. Climate change affects these rights via its effects on food, water and housing through storms, floods, extreme weather and rising sea levels, all of which will further create forced displacement and migration, or, in extreme cases, such as that of Kiribati, relocation of entire communities.

Crimes against humanity

There is no single comprehensive definition of what constitutes crimes against humanity in international law, despite trials having taken place for such crimes. The most widely accepted definition is that adopted by the ICC, although the court itself states that the definition is non-exhaustive and that its scope is for the purpose of the Rome Statute (i.e., states under the jurisdiction of the ICC are bound by the definition). According to the Rome Statute (UN 1998: Article 7/1), which created the ICC, a crime against humanity includes the following acts being committed as part of a widespread or systematic attack directed against any civilian population: murder, extermination, deportation or forcible transfer of population, torture, other inhumane acts of a similar character intentionally causing great suffering, or serious bodily or mental injury. According to the statute, extermination is defined as a crime against humanity when committed as part of a widespread or systematic attack directed against a civilian population. This definition and interpretation of extermination also includes the criterion of 'intentional infliction of conditions of life ... calculated to bring about the destruction of part of a population'. Likewise, the crime of genocide, as defined under Article 6 of the statute, includes 'deliberately inflicting ... conditions of life calculated to bring about its [the targeted groups'] physical destruction in whole or in part' (ICC 2002: 9). Significantly, and suggesting that these definitions are indeed applicable to the concept of humanitarian climate crimes, the statute further defines the denial of 'conditions of life' as including any 'deliberate depravation of resources indispensable for survival, such as food or medical services, or systematic expulsion from homes' (Schabas 2001: 250).

A key point in defining and acting upon humanitarian climate crimes, which I turn to in the following section, is the consideration of awareness as intent. Knowledge of the consequences of an action, particularly when that knowledge is that the actions will be severely harmful, must be interpreted as deliberate, or at minimum as a crime of negligence, despite the consequences not being the primary purpose of the causal action. Therefore, any of the effects of climate change preventing fulfilment of environmentally affected human rights may be considered as deliberate deprivation of resources indispensable for survival, and any effect which necessitates forced migration or displacement may be considered as systematic expulsion from homes. A further relevant provision of current international law in relation to the crime of torture is Article 14 of the Second Additional Protocol to the Geneva Convention (1977). The protocol prohibits the ruin of resources indispensable to the survival of the civilian population, and most relevantly, given the noted indirectness of climate change, prohibits harm to the life and person of citizens, regardless of the methods used to cause it. It follows from this that the destruction wreaked on the human security of future generations, through the method of contaminating air, soil and water, or creating an environment which cannot provide an adequate standard of living, is therefore recognizable as a crime.

To illustrate the plausibility of viewing the causal actions of climate change as humanitarian climate crimes, it is also necessary to recognize that the increased 'natural disasters' it will create are in fact human-made disasters. Although it was a retrospective case rather than a pre-emptive one, the referral (under UN Security Council Resolution 1593) of the 2005 Darfur famine case to the ICC provides a precedent for states being held accountable for criminal behaviour leading to an event which is traditionally viewed as being a natural disaster. It illustrates that what has been generally defined as a natural disaster can instead be recognized as 'a slow indictable crime against humanity' (Edkins 2007: 50). Many climate cases could be mapped onto this model of accountability for causing harm and violating rights via the environment. Submersion of Kiribati due to climate change would be no more a natural disaster than if a weapon was deployed to cause the same harms, so it ought to be treated as an environmental act of aggression. The only natural aspect to these climate-change disasters is the indirect use of the natural environment as a weapon. Inaction to halt or hinder the causal actions of climate change with immediacy sees these looming climate harms being made to happen. Natural disasters caused by climate change do not just happen; they are made to happen and they are committed, which makes them criminal rather than natural (Edkins 2007: 57).

The various injustices which will occur as a direct result of climate change may be framed as any of these grave crimes, based on the application and interpretation of existing definitions. These harms, when they manifest, whether as rights violations, humanitarian threats of 'natural disasters' or crimes against humanity, will also qualify them as public harms because they will be widespread, affecting large collectives as well as individuals. Since there are precedents for famine crimes and mass starvations as being actionable under international law (Edkins 2002, Marcus 2003, Howe and Devereux 2004), then it stands to reason that the wilful causes of such crimes are also actionable. If comparable events which severely harm human security, such as those usually considered to be natural disasters, are caused by wilful actions and treated as crimes, this sets a precedent for a pre-emptive approach to the causal actions of climate change, whereby the consequences qualify as humanitarian climate crimes by causing a permanent or long-ranging ecological change which will have a continuous effect on human security. It seems fair to deduce that causing climate change, or allowing it to be caused, constitutes destruction of the necessary components of subsistence rights, the components for survival and livelihood.

Awareness as intent

If causing climate change is to be viewed as a humanitarian climate crime requiring pre-emptive action, it must be defensible that an actor should be held accountable pre-emptively for a harm that not only has not yet happened, but which they also do not directly intend to cause. Presuming that the severity of the known forthcoming harms, when they occur, qualify them as humanitarian climate crimes, and presuming there will be victims, the extent to which the causal actors wilfully intend to cause the harmful consequences must be considered. Climate change happens via an identifiable chain of cause-and-effect, a chain which is currently known to the state actors perpetrating the harmful behaviours. It is the dual impact of this certainty and current awareness of the climate harms which justifies the causal actions being pre-emptively treated as humanitarian climate crimes in advance of the consequential injustice actually occurring.

When identifying the responsibility of actors and assigning blame, knowledge of the outcome(s) of an action can be equated to deliberate intent. To recognize this awareness as intent it is also presumed that having knowledge of the outcome of an action must include knowledge of the probability of that outcome, which in this case is definite (if causal behaviour and actions continue as they are). For example, in the case of Kiribati becoming submerged, the actors who are responsible for the rising sea levels are aware that Kiribati is predicted to 'sink' before the end of this century. No state or corporate actor is declaring that the purpose of direct intent of their action(s) is to cause a rise in sea levels which submerges an entire state and in turn causes a humanitarian disaster, mass rights violation or crime against humanity. Rather, the primary purpose in a simplified sample chain of events which will bring about a rise in sea level is to bring about employment or to provide energy (action X) for a state's (A) own population (effect Y) and is not undertaken to bring about the submersion of Kiribati (effect Z). However, the knowledge that Z will definitely follow from X, just as certainly as Y will do so, implies a complete willingness to bring about both effects Y and Z by carrying out X. Whether Z occurred or not, A would carry out X to achieve Y. Therefore, by intentionally executing X,

A intends for all of its known outcomes to occur, thus intending both Y and Z. By willingly carrying out an action, all known effects are the intent(s) of that action whether or not they are the primary purpose of the action (Meiland 1970: 7–15). These humanitarian climate crimes which are consequential to climate change are accepted as being significant effects, and if Z is a significant effect of X, the knowledge or awareness of that, combined with the deliberate intent to do X (as opposed to allowing another actor already carrying out action X to continue to do so unimpeded, or A carrying out X by mistake) renders A blameworthy.

It is unconvincing for any actor to claim a lack of awareness of their actions in relation to climate change. As noted previously, this problem is currently acknowledged and recognized as an urgent global justice and human security issue by all UN member states. Although some member states have yet to ratify the Paris Agreement and others plan to withdraw from the Paris Agreement and UNFCCC, all current member states have adopted the (non-binding) 2030 agenda for sustainable development, including the 13th sustainable development goal of Climate Action, indicating the awareness of member states of the gravity of climate change. Although causing a secondary but known effect (Z) may be viewed as negligent or reckless, or even claimed as accidental, ultimately, with regard to the responsibilities assigned, lack of positive preventative actions by certain actors could be considered akin to intent to cause the actual harm by omission. By allowing something (in this case effect Z) to happen, one can be held accountable for what happens, therefore the specific actors who make (or allow) the causal actions to happen should be held equally to blame as if they were deliberately carrying out the actions with the primary purpose of causing the humanitarian climate crime of Z. The humanitarian climate crimes are not necessarily caused by harmful conduct in the strictest ethical sense, as no single actor is maliciously acting with wilful intent to cause the climate injustice single-handedly. Nevertheless, despite this lack of primary intent to cause these harmful consequences, the awareness of the effects must be considered akin to intent in the criminal, legal and moral sense, as the causal actors know what the effects will be.

The affect that intent has on an actor's responsibility for a causal action is varied and complex. In order to be held culpable or accountable it must be just to hold an actor responsible for the effects of their actions. This holds true even in the presence of mitigating circumstances or long chains of cause-and-effect, as in the case of climate change and humanitarian climate crimes. To illustrate the difference between intent, recklessness and negligence, consider a state's (A) harmful emissions levels. If A deliberately emits an amount of greenhouse gases (GHGs) (action X) with at least one of the primary purposes being to cause climate change in order to submerge Kiribati (committing a crime against humanity and violating various human rights en masse in the process), that action equals direct and deliberate intent. However, if state A is aware of the likelihood of causing climate change and is aware of the consequences for Kiribati, and is also aware of any obligations which it will risk not meeting if it continues with action X, yet A still does not lower its emissions levels as required, this qualifies as reckless behaviour or negligence rather than direct intent. Nevertheless, this lack of deliberate desire to cause harm does not excuse the actual causing of harm, particularly where there is awareness of severely harmful consequences. Rather, the prior knowledge of the harmfulness of an action should be considered akin to intent as A carries out X in spite of knowing it will cause effect Z. Regardless of the primary purpose of an action, if the actor is aware of the future harm that will be caused by their action, then that actor is accountable for that harm.

Deliberately carrying out an action which will cause a mass human rights violation or a crime against humanity, even if the action is being carried out for an alternative purpose (e.g. supplying heat to a population via fossil fuel energy) does not exclude the actor from blame, accountability or responsibility. Given the bioregional transience of the causal factors of climate change and its harms, pre-emptive intervention on behalf of the citizens of an area identified as being consequentially harmed in one place (such as Kiribati) should occur to prevent the causal action from taking place. The obligation to pre-emptively prevent harm in the case of climate injustice extends beyond the harm principle (Mill 1956: 13) of refraining from actively doing harm; it also invokes the duty of actively doing good. Committing a harm by omission as opposed to commission is still a wrong, and taking no obligatory action against known forthcoming humanitarian harms is still taking the action of inaction (Minh-ha 1989: 44), which can be considered a crime in itself.

Conclusion

The concept of 'humanitarian climate crime' introduced herein aims to collectively encompass the looming injustices that climate change will cause as it continues to manifest itself with increasing severity. As used in this chapter to describe climate change-caused (or -affected) crimes against humanity or mass human rights violations, the concept is a catch-all concept indicating either an ethical crime (an actual legal crime) or a moral crime (an injustice) severe enough to cause a humanitarian disaster, or a harm necessitating humanitarian aid or intervention. This includes future harms which may, as illustrated, fall under the distinction of a crime against humanity or mass human rights violation, creating a breach of global justice. Given that all state actors are aware of the effects of climate change, those actions can be paralleled with deliberate crimes, as demonstrated in the way that 'natural' disasters can justly be criminalized. By mapping the identified consequences of climate change, exemplified by the case of Kiribati, onto the current frameworks of crimes against humanity and human rights law, causing those consequences are humanitarian climate crimes and so are owed the same preventive protection and intervention that current crimes against humanity and mass human rights violations receive. This alternative framing offers a reason why there are responsibilities to pre-emptively protect future victims from looming harms before they become actual harms.

This alternative framework may be viewed as controversial, and is not without practical barriers to implementation, namely the power imbalances in the global system despite legal sovereign equality, the lack of political will to take actions against other states or the unwillingness of states to make the necessary changes themselves, the fact that not all states are under the jurisdiction of the ICC, and the fact that most developed states share the bulk of blame. Nevertheless, the concept of humanitarian climate crime offers a useable framework through which to attempt to change the narrative of climate change and view it through a lens of criminal intent rather than to see it as an abstract and unavoidable by-product of current global practices. There are a steadily increasing number of climate cases being taken by citizens against states to hold them to account for not meeting their obligations to protect current citizens and/or future generations from climate change. These cases set precedents for state actors being held accountable for their causal behaviour, which in turn could see citizens in one state take a case against a third-party state, and states taking cases against other states to enable them to meet their own obligations. Furthermore, on paper there is ample evidence to interpret the causal actions of climate change as deliberate harm under the mechanisms explicated above, and the threat of this being put into practice could function as a disciplinary power which could itself act as a deterrent to actors committing humanitarian climate crimes or be an incentive to meet obligations to protect against them (Foucault 1994: 57–8).

In spite of evidence that climate change increases occurrences of 'natural disasters' (IPCC 2013, Field 2012, Seneviratne et al. 2012), many of these disasters and the subsequent harms are anthropogenic in their cause and thus are unnatural. To further the fight for climate justice for future generations, it is essential to take enforceable effective action immediately. Reframing the causal actions for climate change as intentional humanitarian climate crimes that can justifiably be tackled pre-emptively means that enforceable action to stop harmful causal behaviour can be taken based on forthcoming known outcomes. This changes the approach to tackling climate change, from being an aspirational and voluntary target to something which is obligatory for state actors and which can be tackled immediately, without the delay that the creation of new systemic legal or political instruments or mechanisms would require. As the battle for climate justice is already extremely urgent, framing the impacts of climate change in terms of humanitarian climate crime allows action to be taken immediately, rather than awaiting an ever-postponed international consensus to do so. This may give the best possible chance of achieving climate justice for future generations.

References

BBC (2018), 'Japan heatwave declared natural disaster as death toll mounts', accessed 20 May 2019 at: https://www.bbc.com/news/world-asia-44935152.

Collins-Chobanian, Shari (2000), 'Beyond Sax and welfare interests', *Environmental Ethics*, **22**(2), 133–48.

Council of Europe (1950), 'European Convention for the Protection of Human Rights and Fundamental Freedom', Strasbourg: Council of Europe, accessed 11 March 2019 at: www.echr.coe.int/Documents/Convention_ENG.pdf.

Edkins, Jenny (2002), 'Mass starvations and the limitations of famine theorising', *IDS Bulletin*, **33**(4), 12–18.

Edkins, Jenny (2007), 'The criminalization of mass starvations: from natural disaster to crime against humanity', in Stephen Devereux (ed.), *The New Famines: Why Famines Persist in an Era of Globalization*, Oxford: Routledge, pp. 50–65.

Evans, Graham and Jeffrey Newnham (1998), *The Penguin Dictionary of International Relations*, London: Penguin.

Field, Christopher B. (ed.) (2012), *Managing the Risks of Extreme Events and Disasters to Advance Climate Change Adaptation: Special Report of the Intergovernmental Panel on Climate Change*, Cambridge: Cambridge University Press.

Foucault, Michel (1994), 'Truth and juridical forms', in James D. Faubion (ed.), *Power: Essential Works of Foucault 1954–1984*, New York, The New Press, pp. 1–89.

Francioni, Francesco (2010), 'International human rights in an environmental horizon', *The European Journal of International Law*, **21**, 41–55.

Geneva Convention (1949), 'Relative to the protection of civilian persons in time of war' (Fourth Geneva Convention), International Committee of the Red Cross (ICRC) Geneva: 12 August (Second Additional Protocol added 1977).

Howe, Paul and Stephen Devereux (2004), 'Famine intensity and magnitude scales: a proposal for an instrumental definition of famine', *Disasters*, **28**(4), 353–72.

ICISS (2001), 'The responsibility to protect', Gareth Evans and Mohamed Sahnoun (co-chairs), *Report of the International Commission on Intervention and State Sovereignty*, Ottawa: International Development Research Centre.

Intergovernmental Panel on Climate Change (IPCC) (2013), 'Climate change 2013: the physical science basis', Contribution of Working Group 1 to the Fifth Assessment Report of the Intergovernmental Panel on Climate Change, Thomas Stocker, Qin Dahe, Gian-Kaspar Plattner, M. Tignor, S.K. Allen, J. Boschung, A. Nauels et al. (eds), Cambridge: Cambridge University Press.

Intergovernmental Panel on Climate Change (IPCC) (2018), 'Global warming of 1.5°C. An IPCC Special Report on the impacts of global warming of 1.5°C above pre-industrial levels and related global greenhouse gas emission pathways, in the context of strengthening the global response to the threat of climate change, sustainable development, and efforts to eradicate poverty', V. Masson-Delmotte, P. Zhai, H-O. Pörtner, D. Roberts, J. Skea, P.R. Shukla, A. Pirani et al. (eds), Geneva: World Meteorological Organization.

International Criminal Court (ICC) (2002), 'Rome Statute of the International Criminal Court', The Hague: ICC Public Information and Documentation Centre, accessed 25 January 2018 at: www.icc.cpi.int.

Lazzari, Agnese, Chiara De Waure and Natasha Azzopardi-Muscat (2015), 'Health in all policies', in Stefania Boccia, Paolo Villari and Walter Riccardi (eds), *A Systematic Review of Key Issues in Public Health*, New York: Springer International Publishing, pp. 277–86.

Linklater, Andrew (2005), 'The English School', in Scott Burchill, Andrew Linklater et al. (eds), *Theories of International Relations*, Basingstoke: Palgrave Macmillan, pp. 84–109.

Maastricht Treaty (1992), 'Treaty on European Union' (Consolidated Version), Treaty of Maastricht, *Official Journal of the European Communities*, accessed 24 November 2018 at: http://www.refworld.org/docid/3ae6b39218.html.

Mahendra, Pal, Ayele Yodit, Angesom Hadush, Sumitra Panigrahi and Vijay J. Jadhav (2018), 'Public health hazards due to unsafe drinking water', *Air and Water Borne Diseases*, **7**(1) accessed 8 October 2019 at https://www.omicsonline.org/open-access/public-health-hazards-due-to-unsafe-drinking-water-2167-7719-1000138-101933.html.

Marcus, David (2003), 'Famine crimes in international law', *American Journal of International Law*, **97**, 245–81.

McCusker, Kelly E., David S. Battisti and Cecilia M. Bitz (2015), 'Inability of stratospheric sulfate aerosol injections to preserve the West Antarctic ice sheet', *Geophysical Research Letters*, **42**(12), 4989–97.

Meiland, Jack W. (1970), *The Nature of Intention*, London: Methuen and Co.

Mill, John S. (1956), *On Liberty*, Currin V. Shields (ed.), Indianapolis, IN: Bobbs-Merrill Co.

Minh-ha, Trin T. (1989), *Woman, Native, Other*, Indianapolis, IN: Indiana University Press.

Nickel, James W. (1993), 'The human right to a safe environment: philosophical perspectives on its scope and justification', *Yale Journal of International Law*, **18**, 281–95.

O'Doherty, Selina (2018), 'Environmental human rights: concepts of responsibility', in Markku Oksanen, Ashley Dodsworth and Selina O'Doherty (eds), *Environmental Human Rights: A Political Theory Perspective*, Oxford: Routledge, pp. 149–66.

OHCHR (2009), 'Annual report of the United Nations High Commissioner for Human Rights and Reports of the Office of the High Commissioner and the Secretary General', accessed 28 January 2019 at: https://documents-dds-ny.un.org/doc/UNDOC/GEN/G09/103/44/PDF/G0910344.pdf.

Owens, Patricia (2012), 'Human security and the rise of the social', *Review of International Studies*, **38**(3), 547–67.

Page, Edward (2007), *Climate Justice and Future Generations*, Cheltenham, UK and Northampton, MA, USA: Edward Elgar Publishing.

Palmer, Ewan (2018), 'Heat wave continues across US as death toll climbs', *Newsweek*, accessed 20 May 2019 at: https://www.newsweek.com/heat-wave-continues-across-us-death-toll-climbs-1008162.

Public Health England (PHE) (2019), 'PHE heatwave mortality monitoring Summer 2018', London: Public Health England, accessed 20 May 2019 at: https://assets.publishing.service.gov.uk/government/uploads/system/uploads/attachment_data/file/771819/PHE_heatwave_mortality_monitoring_report_2018.pdf.

Ramnewash-Oemrawsingh, Sangini (2011), *The Human Right to a Viable Environment*, The Hague: Asser Press.

Samson, J., D. Berteaux, B.J. McGill and M.M. Humphries (2011), 'Geographic disparities and moral hazards in the predicted impacts of climate change on human populations', *Global Ecology and Biogeography*, **20**, 532–44.

Sax, Joseph L. (1990), 'The search for environmental rights', *Journal of Land Use and Environmental Law*, **6**, 92–105.

Schabas, William A. (2001), *An Introduction to the International Criminal Court*, Cambridge: Cambridge University Press.

Seneviratne, Sonia I., Neville Nicholls, David Easterling, Claire M. Goodess, Shinkiro Kanae, James Kossin and Yali Luo (2012), 'Changes in climate extremes and their impacts on the natural physical environment', in Christopher B. Field (ed.), *Managing the Risks of Extreme Events and Disasters to Advance Climate Change Adaptation: Special Report of the Intergovernmental Panel on Climate Change*, Cambridge: Cambridge University Press, pp. 109–230.

Shelton, Dinah L. (2006), 'Human rights and the environment', *Denver Journal of International Law and Policy*, **35**, 129–71.

Storey, Donovan and Shawn Hunter (2010), 'Kiribati: an environmental "perfect storm"', *Australian Geographer*, **41**(2), 167–81.

United Nations (UN) (1948), 'Universal Declaration of Human Rights', Paris: UN General Assembly, accessed 22 July 2018 at: www.ohchr.org/Documents/Publications/Compilation1.1en.pdf.

United Nations (UN) (1966), 'International Covenant on Economic, Social and Cultural Rights', New York: UN General Assembly, accessed 11 March 2019 at: http://www.ohchr.org/EN/Professional Interest/Pages/CESCR.aspx.

United Nations (UN) (1998 [last amended 2010]), 'Rome Statute of the International Criminal Court', Rome: UN General Assembly and Kampala Review Conference, accessed 8 May 2018 at: http://www.refworld.org/docid/3ae6b3a84.html.

Vaha, Milla E. (2015), 'Drowning under: small island states and the right to exist', *Journal of International Political Theory*, **11**(2), 206–23.

World Health Organization (WHO) (2013), 'The right to health', Geneva: UN Office of the High Commissioner for Human Rights, accessed 12 November 2018 at: http://www.who.int/mediacentre/factsheets/fs323/en/.

12 Climate justice after the Paris Agreement: understanding equity through nationally determined contributions

Claire Swingle

To better understand how Parties to the Paris Agreement on climate change conceptualize climate justice, this chapter analyses the 'Fairness and Ambition' sections within 163 Nationally Determined Contributions (NDCs) submitted to the secretariat of the United Nations Framework Convention on Climate Change (UNFCCC or Convention hereafter). Each Party's indicators of equity are compared against the traditional positions of their respective negotiating groups and against more recent submissions to the UNFCCC under the Talanoa Dialogue (TD), the facilitative dialogues convened throughout 2018 in order to take stock of the collective efforts of Parties and to inform the preparation of updated NDCs by 2020.

This analysis reveals a disconnect between the positions that Parties and coalitions have taken in their NDC and TD submissions, on one hand, and the positions that they are most known for promoting, as groups, in the UNFCCC negotiations, on the other hand. Specifically, historical responsibility is underemphasized and decoupled from common but differentiated responsibility (CBDR), which had often been invoked by developing Parties as a call for developed Parties to take stronger action. Instead, temperature goals and references to scientific reports are increasingly used to build calls for enhanced ambition and international support.

How Parties conceptualize equity will affect how they judge global mitigation action and shape their subsequent goals. This will be crucial to limiting global warming – the most pressing issue of climate justice – because all Paris Agreement commitments are voluntary and because Parties' ambition to address climate change needs to be quickly increased. In this chapter I describe a framework through which climate justice can be better understood as a post-Paris climate regime emerges and evolves. I highlight several ways in which further research is needed.

Background: equity in Nationally Determined Contributions

Debates about who should do what – fundamentally questions of equity – have been central to the international climate regime for decades. The Convention recognizes 'common but differentiated responsibilities and respective capabilities' in acting on climate change as one of its guiding principles. In so doing, it explicitly recognizes a global duty to address climate change, but that the extent to which different Parties are expected to act should not be the same. This differentiation in responsibility and capability to act, captured by the phrase CBDR, is the form that equity has assumed within the international climate regime over the last two decades. However, the understanding and application of CBDR itself has shifted from the strict differentiation in commitments reflected in the Kyoto Protocol toward increasingly expecting all Parties to make emissions reductions, albeit to varying degrees and while recognizing their development priorities (Rajamani 2012). The Paris Agreement continues this trend by requiring each Party to submit a Nationally Determined Contribution toward achieving the objectives of the agreement, in the process allowing each of them to decide what commitments it will make.

This chapter adds to existing literature on the equity implications of Paris commitments, focusing specifically on the content of equity claims made by Parties themselves, something that has been missing from many analyses hitherto. For instance, The German Development Institute's 'NDC Explorer' provides information only as to whether or not a section on fairness is included in a Party's submission (German Development Institute 2019). The Civil Society Review (2015) attempts to judge whether proposed efforts within NDCs are fair. However, it does not consider Parties' own conceptions of equity and fairness, instead imposing a carbon budget framework that is divided by responsibility and capability.

The UNFCCC secretariat's 2016 synthesis report on NDCs (UNFCCC secretariat 2016) is the only document to date that has summarized what Parties consider to be equitable in their submissions. This is a step in the right direction, but it dedicates only one of 75 pages to this avenue of inquiry, and it does not provide any systematic way to analyse the contributions going forward. The report merely lists indicators that many Parties included in their NDCs, without identifying which Party said what, whether that was consistent or a change from previous positions, or how that relates to overall ambition level. It also states that '*All* Parties included a narrative on how they consider their [NDCs] to be fair and ambitious. . .' (UNFCCC secretariat 2016: 8; emphasis added). However, my analysis finds that 20 Parties did not include any such narrative. Without providing any methodology for their synthesis, we cannot assess what assumptions the secretariat may have made to reach its conclusions. There are, likewise, several synthesis reports on Talanoa Dialogue submissions (UNFCCC secretariat 2018a, 2018b), but they, too, show only high-level trends and do not analyse which Parties have taken which positions and whether or not those are consistent with positions put forward in NDCs.

This is a significant shortcoming. Since climate change is a problem of the commons with classic collective action and free-rider problems, it is essential that Parties see that others are doing their 'fair share' if they can be expected fully to act themselves. For instance, China's submission to the Talanoa Dialogue states that 'Only an equitable mechanism can attract extensive and universal participation, build mutual trust, and inspire concrete actions. It is one of the most crucial elements for the success in combating climate change' (China 2018: 5–6).

Given that the post-Paris international climate regime relies on the voluntary contributions of Parties, it is imperative to understand how each Party conceptualizes its own fair share, and that of other Parties. It is especially urgent to do so because Parties must quickly ratchet up their ambition, as current NDCs, even if fully implemented, will not limit global warming to 2°C, let alone the 1.5°C that is aspired to in the agreement (UNEP 2018). Moreover, without strong enforcement mechanisms, implementation of, and compliance with, the Paris Agreement relies on transparency and processes driven by mutual support and capacity building, and potentially on naming and shaming. A clear understanding of Party positions facilitates this transparency.

Methodology

To address the gap in existing analyses, I analysed each NDC submitted to the UNFCCC online portal and compiled information on indicators of equity that each Party considered. The resulting data were collected into one database (which is available in full at https://paulgharris.files.wordpress.com/2019/04/ndc-database-c.-swingle.pdf). I considered all NDCs, irrespective of whether the submitting country has ratified the Paris Agreement or not. I did not attempt to resolve the question of whether a Party's NDC is fair.

The Lima Call for Climate Action provided guidelines as to what each NDC should include, and it stated that a Party *may* include a section on 'how the Party considers that its intended nationally determined contribution is fair and ambitious' (UNFCCC secretariat 2014: 2). Although voluntary, 136 of the 163 Parties that submitted an NDC as of January 2018 included this specific section, and seven others included explicit statements about what they view as fair and ambitious. Below I analyse only the NDC section on fairness and ambition. Because each NDC is self-generated, it is presumably seen as generally fair to the Party that presents it; nonetheless, an explicit explanation of the Party's conceptualization of fairness is important. The Lima Call for Climate Action outlined the same opportunity to every Party to clearly articulate its views on equity, so even the decision to include or not to include this section is significant. Moreover, while similar elements may be referenced throughout an NDC, Parties might view them as contributing differently to equity, which could be made clear in this section. There could also be a gap in what a Party views as fair and what it is actually proposing to undertake, due to capability restraints or inaction by others.

Here I use 'equity' and 'fairness' according to common usage. In official documents, 'fairness' is often used in conjunction with 'ambition', stemming from the above-mentioned language in the Lima Call for Climate Action, while 'equity' is normatively used in other contexts relating to both procedural and distributive equity, for example: 'The aggregate consideration process shall be conducted consistent with science and on the basis of equity...' (UNFCCC secretariat 2015: 76). This is also the approach adopted by the IPCC in its fifth Assessment Report (Fleurbaey et al. 2014: 294).

My analysis centres primarily around negotiating groups, which represents just one potential use of the dataset. Negotiating groups are a prominent structure in the UNFCCC. Understanding trends in how they conceptualize equity could be useful for increasing the ambition of their commitments. It is also interesting that the NDCs were prepared individually and yet Parties are still part of collective negotiating groups at the UNFCCC. I analysed equity indicators included in NDCs between these groups to see how equity perceptions compare. The number of individual Parties taking a position within a negotiating group divided by the number of Parties within that group was compared to the 'average of all Parties', calculated by dividing the number of Parties that included that indicator by 143, the number of NDCs containing a section on fairness and ambition. It is not uncommon for a Party to be a member of several negotiating groups concurrently. In such cases, those Parties' submissions were counted each time, creating a bit of bias toward the positions of developing countries, which are more likely than developed countries to be part of multiple negotiating groups.

I performed multiple regression analysis of the dataset to understand relationships between the inclusion of different indicators of equity and negotiating group membership, using the model $Y = \beta_0 + \beta_1 X_1 + \beta_2 X_2 + \beta_n X_n + \varepsilon$, where Y is the indicator of equity (i.e., historical responsibility, science, etc.), β_0 is the intercept, $X_1... X_n$ are predictor variables (i.e., membership in different negotiating groups), controlling for change in global emissions from 1990–2014, and ε is the error term. The R^2 values for each regression were quite low at 15–20 per cent, indicating that the predictor variables explain only part of the variance in the outcome variable (whether a certain equity indicator was included), which is to be expected given that a Party's decision to include an indicator is quite complicated. Nonetheless, the relationships between certain variables are statistically significant and point out important trends.

I also compared indicators of equity considered in Talanoa Dialogue documents submitted to the official UNFCCC portal between 1 March and 31 December 2018 against those in NDCs to understand trends and shifts in official conceptions of climate justice. Talanoa Dialogue submissions are slightly different from NDCs because they are part of a stocktaking exercise and are not legally binding. There was also no explicit recommendation to include a section on fairness and ambition, as there was for NDCs. Consequently, the entire submissions were considered rather than solely the sections on equity. Furthermore, Parties were able to submit individually or as a group of several Parties, or both, unlike with the NDCs, where each Party was required to submit individually. Still, TD submissions are the most recent documents that can be used to understand official Party positions. They provide valuable information on perceptions of climate justice.

Findings on equity conceptualizations in NDCs

Do the criteria for equity included in NDCs create 'self-serving' equity claims (Lange et al. 2010)? Do they align with negotiating coalitions' traditional positions? Are they the 'core equity principles' that the IPCC has recognized as serving 'the basis for most discussions of equitable burden sharing in a climate regime', namely historical responsibility, capability, development need and equality (Fleurbaey et al. 2014: 287)?

Summary statistics

Analysis of the 163 submissions yielded 16 primary indicators considered by Parties in their conceptions of equity: historical emissions responsibility (aggregate and per capita); current emissions responsibility (aggregate and per capita); projected emissions responsibility (aggregate and per capita); national capabilities; carbon intensity; cost effectiveness; past efforts; low responsibility; specific temperature goal; scientific reports; adaptation; international support; and CBDR. Figure 12.1 provides an overview of the frequency at which each was considered.

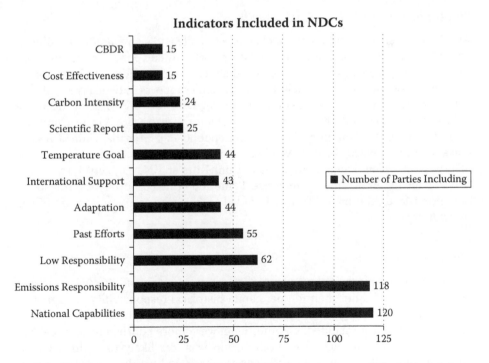

Note: For emissions responsibility, 34 Parties included historical emissions responsibility, 143 included current emissions responsibility, and 36 included projected emissions responsibility. For the temperature goal, seven referenced 1.5°C, four 1.5–2°C, 33 2°C and one above 2°C.

Figure 12.1   Equity indicators included in NDCs

Negotiating coalitions

The African Group

In comparison to the average of all Parties, the African Group placed less emphasis on historical responsibility (statistically significant with coefficient of −0.23 and p-value of 0.010), carbon intensity, cost effectiveness and how its NDC represents a step up from past efforts. Greater than average emphasis was placed on acting despite low responsibility and the importance of international support. Data for the African Group came from the NDCs of Algeria, Angola, Benin, Botswana, Burkina Faso, Burundi, Cameroon, Cape Verde, Central African Republic, Chad, Comoros, Democratic Republic of Congo, Republic of Congo, Cote D'Ivoire, Djibouti, Egypt, Equatorial Guinea, Eritrea, Ethiopia, Gabon, Gambia, Ghana, Guinea-Bissau, Guinea, Kenya, Lesotho, Liberia, Madagascar, Malawi, Maldives, Mali, Mauritania, Mauritius, Morocco, Mozambique, Namibia, Niger, Nigeria, Rwanda, Sao Tome and Principe, Senegal, Seychelles, Sierra Leone, Somalia, South Africa, South Sudan, Sudan, Swaziland, Tanzania, Togo, Tunisia, Uganda, Zambia and Zimbabwe (UNFCCC secretariat 2019).

The Alliance of Small Island States (AOSIS)

AOSIS members more frequently included per capita and aggregate responsibility, national capabilities, 1.5°C as the temperature goal, why current commitments represent a step up from past efforts, and how they are acting despite low responsibility. They were also more likely to include international support (statistically significant with coefficient of 0.350 and p-value of 0.010). Data for AOSIS came from the NDCs of Antigua and Barbuda, Bahamas, Barbados, Belize, Cape Verde, Comoros, Cook Islands, Cuba, Dominica, Dominican Republic, Fiji, Grenada, Guinea-Bissau, Guyana, Haiti, Jamaica, Kiribati, Maldives, Marshall Islands, Mauritius, Micronesia, Nauru, Niue, Palau, Papua New Guinea, Saint Kitts and Nevis, Saint Lucia, Saint Vincent and the Grenadines, Samoa, Sao Tome and Principe, Seychelles, Singapore, Solomon Islands, Suriname, Tonga, Trinidad and Tobago and Vanuatu (UNFCCC secretariat 2019).

The Arab Group

The Arab Group placed below-average emphasis on responsibility (past, current and future) and did not mention per capita historical responsibility or aggregate projected responsibility. There was also less mention of international support. However, when it is included, commitments were more contingent on receiving that support than average. The Arab Group was less likely to include CBDR (statistically significant with a coefficient of −0.161 and p-value of 0.049), a temperature goal or science national capabilities. There is above average emphasis on adaptation and why past efforts justify inaction presently. Data for the Arab Group came from the NDCs of Algeria, Bahrain, Comoros, Djibouti, Egypt, Iraq, Jordan, Kuwait, Lebanon, Mauritania, Morocco, Oman, Qatar, Saudi Arabia,

Somalia, Sudan, Tunisia, United Arab Emirates and Yemen (UNFCCC secretariat 2019).

The Brazil, South Africa, India, China Group (BASIC)

The BASIC countries emphasized historical responsibilities, especially per capita (statistically significant with a coefficient of 0.286 and p-value of 0.009), national capabilities, greenhouse gas (GHG) intensity (statistically significant with a coefficient of 0.367 and p-value of 0.052), past efforts, CBDR and international support for adaptation. The BASIC group places less emphasis on aggregate current responsibility and does not mention aggregate projected responsibility (UNFCCC secretariat 2019).

The Environmental Integrity Group (EIG) and the Umbrella Group

The EIG/Umbrella Group places above average emphasis on per capita current responsibility and per capita and aggregate projected responsibility, carbon intensity, the temperature goal (only 2°C) and past efforts. They mention historical responsibility and adaptation at about the average rate. They place below average emphasis on international support. They are more likely to include cost effectiveness (statistically significant with a coefficient of 0.256 and p-value of 0.014) and reference to scientific reports (statistically significant with a coefficient of 0.859 and p-value of 0.0003). Data for the EIG and Umbrella Group came from the NDCs of Australia, Canada, Iceland, Japan, Liechtenstein, Mexico, New Zealand, Norway, Republic of Korea, Russia, Switzerland, Ukraine and the United States (UNFCCC secretariat 2019).

European Union (EU)

The EU places below average emphasis on historical responsibility, how low responsibility affects action, CBDR, adaptation international support. It is about average in considering current and projected responsibility, although per capita consideration is slightly higher. It places above average emphasis on GHG intensity (statistically significant with a coefficient of 0.890 and p-value of 0.014) and past efforts (statistically significant with a coefficient of 2.199 and p-value of 0.021). Data for the EU came from the organization's collective NDC (UNFCCC secretariat 2019).

The Group of 77 (G77) and China

The use of indicators by the combined group of G77 developing countries and China tracks closely that of all Parties, perhaps reflecting that the G77 is the largest coalition. They do slightly underemphasize responsibility, especially projected responsibility (statistically significant with a coefficient of −0.118 and p-value of 0.0337), cost effectiveness, GHG intensity and past efforts. They slightly overemphasize how low responsibility affects action, adaptation, CBDR and international support. Data for the G77 and China came from the NDCs of Afghanistan,

Algeria, Angola, Antigua and Barbuda, Argentina, Bahamas, Bahrain, Bangladesh, Barbados, Belize, Benin, Bhutan, Bolivia, Bosnia and Herzegovina, Botswana, Brazil, Brunei Darussalam, Burkina Faso, Burundi, Cambodia, Cameroon, Cape Verde (revised INDC), Central African Republic, Chad, Chile, China, Colombia, Comoros, Democratic Republic of Congo, Congo, Costa Rica, Cote D'Ivoire, Cuba, Djibouti, Dominica, Dominican Republic, Ecuador, Egypt, El Salvador, Equatorial Guinea, Eritrea, Ethiopia, Fiji, Gabon, Gambia, Ghana, Guatemala, Guinea-Bissau, Guinea, Guyana, Haiti, Honduras, India, Indonesia, Iran, Iraq, Jamaica, Jordan, Kenya, Kuwait, Lao, Lebanon, Lesotho, Liberia, Madagascar, Malawi, Malaysia, Maldives, Mali, Marshall Islands, Mauritania, Mauritius, Micronesia, Mongolia, Morocco, Mozambique, Myanmar, Namibia, Nauru, Nepal, Niger, Nigeria, Oman, Pakistan, Panama, Papua New Guinea, Paraguay, Peru, Philippines, Qatar, Rwanda, Saint Kitts and Nevis, Saint Lucia, Saint Vincent and the Grenadines, Samoa, Sao Tome and Principe, Saudi Arabia, Senegal, Seychelles, Sierra Leone, Singapore, Solomon Islands, Somalia, South Africa, South Sudan, Sri Lanka, Sudan, Suriname, Swaziland, Tajikistan, Tanzania, Thailand, Timor-Leste, Togo, Tonga, Trinidad and Tobago, Tunisia, Turkmenistan, Uganda, United Arab Emirates, Uruguay, Vanuatu, Venezuela, Vietnam, Yemen, Zambia and Zimbabwe (UNFCCC secretariat 2019).

Least Developed Countries (LDCs)

In comparison to the average of all Parties, LDCs more frequently emphasize cost effectiveness (statistically significant with a coefficient of 0.114 and p-value of 0.059), international support (statistically significant with a coefficient of 0.400 and p-value of 0.003), and how their commitment is a step up of projected responsibility (statistically significant with a coefficient of 0.148 and p-value of 0.0002). LDCs less often emphasize past efforts and that low responsibility is a justification for inaction. Data for the LDCs came from the NDCs of Afghanistan, Bangladesh, Benin, Bhutan, Burkina Faso, Burundi, Cambodia, Cape Verde, Chad, Comoros, Djibouti, Eritrea, Ethiopia, Gambia, Guinea-Bissau, Guinea, Haiti, Lao, Malawi, Maldives, Mali, Mauritania, Moldova, Mozambique, Myanmar, Nepal, Niger, Rwanda, Samoa, Sao Tome and Principe, Senegal, Solomon Islands, Somalia, Sudan, Tanzania, Timor-Leste, Togo, Vanuatu, Yemen and Zambia (UNFCCC secretariat 2019).

Like-Minded Developing Countries (LMDCs)

LMDCs place below-average emphasis on per capita historical responsibility (no members mention it) and current responsibility (statistically significant with a coefficient of −0.367 and p-value of 0.006). They less often consider science, the temperature goal (only 2°C), international support, how commitment is a step up from previous efforts, and national capabilities (statistically significant with a coefficient of −0.349 and p-value of 0.032). There are no instances of citing low responsibility as a reason not to act, but they do highlight ambitious past efforts as a reason that the NDC commitment may not need to be ramped up. They emphasize carbon intensity, cost effectiveness and CBDR more than average. Data for the

LMDCs came from the NDCs of Bolivia, China, Cuba, Dominica, Ecuador, Egypt, El Salvador, India, Iran, Iraq, Malaysia, Mali, Pakistan, Philippines, Saudi Arabia, Sri Lanka, Sudan and Venezuela (UNFCCC secretariat 2019).

The Organization of Petroleum Exporting Countries (OPEC)

OPEC places less emphasis on responsibility (past, current and projected) and more emphasis on adaptation and cost effectiveness. There is no mention of international support. Data for OPEC came from the NDCs of Algeria, Ecuador, Iran, Iraq, Kuwait, Nigeria, Qatar, Saudi Arabia and Venezuela (UNFCCC secretariat 2019).

Unexpected findings in NDCs

Responsibility is the first part of 'common but differentiated responsibilities and respective capabilities', and generally refers to a Party's cumulative emissions from a certain date. There is disagreement on whether per capita or aggregate emissions should be considered, with high-population developing countries – notably India – calling for per capita use as a measure of equality, arguing that each person should be allocated equal emissions allowances (BASIC experts 2011). Developed countries have traditionally sought to diminish the importance of historical responsibility, especially of emissions before 1990, arguing that they did not yet know the harmful effects of those emissions and should not, therefore, be held accountable for them (Fleurbaey et al. 2014: 318). Moreover, since developing countries have begun to emit considerably more since the Kyoto Protocol was agreed in 1997, some developed countries argue that historical emissions are less important than current and future emissions (Fleurbaey et al. 2014: 318).

If the sections on 'fairness' and 'ambition' simply created the kinds of claims that one might expect countries to make if those sections served to advance their own national interests, more developed Parties (such as the EIG/Umbrella Group coalition and EU) would focus less on historical per capita emissions and more on future emissions. The opposite would be expected of coalitions with quickly developing Parties, such as BASIC and LMDCs. Meanwhile, least developed groups, such as the African Group, LDCs and AOSIS, would likely emphasize both historical and current responsibility. True to expectations, the EIG/Umbrella Group and EU coalitions do emphasize projected responsibility more than average, and they place less emphasis on historical responsibility, with the EU not mentioning per capita or aggregate historical responsibility. Also true to expectations, BASIC, AOSIS and LDCs emphasize historical responsibilities, (low) national capabilities and international support. The LMDCs highlight how they have already undertaken ambitious past efforts, and consequently that their NDC commitments may not need to be ramped up.

However, even though the EIG/Umbrella Group emphasizes projected emissions more than historical ones, the group still considers historical emissions at about the average of all Parties, even though the average is slightly biased toward developing countries. Additionally, this group more often considers per capita, as opposed to aggregate, measurements of emissions responsibility. The EIG/Umbrella Group also emphasizes the importance of adaptation at about the same proportion as developing countries. This could indicate a nod to the group's responsibility to act beyond domestic mitigation.

It is also unexpected that LDCs, BASIC and OPEC do not invoke low responsibility to justify inaction. The African Group, AOSIS, LDCs and LMDCs emphasize that they are acting despite low responsibility, forming a moral call to action. For instance, Gambia's NDC states that, 'By presenting this [NDC], the Gambia would like to provide a moral voice for all responsible and capable countries to undertake actions that are proportionate for their responsibilities and capabilities not only for themselves, but for the whole global community' (Gambia 2016: 4). The LMDCs, Arab Group and OPEC members do not mention per capita historical emissions and underemphasize international support (as do African Group and BASIC). G77+China and African Group also slightly underemphasize historical responsibility.

As a whole, historical responsibility figures less significantly in NDCs than it has in international discussions hitherto. Statistical analysis likewise finds that the change in proportion of global emissions from 1990 to 2014 is not correlated to inclusion of historical responsibility as an indicator within NDCs, although one might expect that an increase in emissions (indicating a developing country) would be correlated to calls to consider historical rather than current or projected responsibility. Increases in the proportion of global emissions are, however, correlated with calls for international support (coefficient of 0.261 and p-value of 0.041). This begins a trend, which continues through the TD submissions, of developing countries calling for increased adaptation and international support rather than focusing on assigning emissions responsibility.

The NDCs also point to the decoupling of CBDR from historical responsibility in official conceptions of climate justice. In climate negotiations, CBDR has often been invoked by developing Parties as a call for developed Parties to take stronger action due to their historical emissions responsibility and because developing Parties are, generally, more vulnerable and have limited capacities to adapt. So, we might expect CBDR to be correlated with inclusion of historical responsibility. However, while it is developing countries – BASIC, G77+China and LMDCs – that emphasize CBDR in their NDCs, only two Parties that invoke CBDR also include historical responsibility. Likewise, there is no significant correlation between CBDR and national capabilities.

Comparison of NDCs to Talanoa Dialogue submissions

The Talanoa Dialogue, which is the first official stocktaking of the Paris Agreement commitments, provides an opportunity to assess how official positions on climate justice have evolved since Parties submitted their NDCs. Nineteen Parties and 11 groups of Parties made TD submissions. Palestine, which did not submit an NDC, also made a TD submission. Almost every submission recognized that NDCs are not ambitious enough, and will not, even if fully implemented, achieve the long-term goal of the Paris Agreement. Rather than undermining confidence in the process, however, Parties emphasized the need for increased collaboration. One notable change is the rhetoric around the urgency of action, including stronger language used in some submissions calling for developed countries to do more.

Maintaining trends established in NDCs, TD submissions de-emphasized emissions responsibility, decoupled CBDR from historical responsibility, and emphasized adaptation and international support. Explicit mention of equity/fairness, past efforts, low responsibility, national capabilities and responsibility (historical, current projected) were emphasized less than they were in NDCs, while international support, adaptation, CBDR, other UNFCCC commitments, temperature goals and science were emphasized more than in NDCs. In general, this creates a sense of urgency for moving forward, rather than a desire to parse out responsibilities.

The most emphasized indicators in the Talanoa Dialogue

International support, adaptation, CBDR, other UNFCCC commitments, temperature goals and science were emphasized more frequently than they were in NDCs. Parties placed importance in being seen as 'good' international players doing their fair share with respect to climate change action. To this end, two of the most commonly used indicators in TD submissions were references to actions taken since submitting the NDC and references to other UNFCCC commitments. For instance, the EU declares that, 'In conformity with the Doha Amendment, for which the EU and Member States have deposited ratification instruments, our actions have already resulted in exceeding our 2020 target to reduce emissions by 20% from 1990', in order to demonstrate that the EU is meeting international obligations and taking ambitious climate action (European Union 2018: 11). On the other hand, references to UNFCCC commitments were also used to define responsibility and to argue that developed countries need to do more because they are not in compliance with commitments already made.

Temperature goals and scientific reports were also more frequently cited in TD submissions than in NDCs. Those indicators were used primarily to demonstrate vulnerabilities to climate change and to show that existing actions and commitments are insufficient. The ambition of temperature goals increased in TD submissions compared to NDCs. Every Party that included a temperature goal in its TD submission that did not do so in its NDC had a goal of keeping temperature rise below 2°C, with the exception of China, which kept it at 2°C. The EIG/Umbrella Group, Norway and the African Group mention 1.5–2°C goals more frequently. Kenya references 1.5°C in its TD submission, whereas it referenced only 2°C in its NDCs. CARICOM even reframes the commitment made in Paris to say that 'Under the Paris Agreement, all Parties have committed to pursuit of a 1.5°C limitation in global average surface temperature increase above pre-industrial levels' (CARICOM 2018: 6).

International support and adaptation were mentioned more frequently in TD submissions, with the former being the most referenced equity indicator across submissions. These were not exclusively invoked by developing countries. The EU, which did not mention adaptation or international support in its NDC, stressed in the Talanoa Dialogue that it is doing its part to assist developing countries, as did Japan and Australia.

Mentions of CBDR also increased in TD submissions. As with NDCs, CBDR was referenced mostly by developing countries. However, this does not necessarily mean that developing countries invoked CBDR so as to implore developed countries to act more, as it has often been employed hitherto. As in NDCs, TD submissions that included CBDR often did not mention historical responsibility; of the ten submissions with CBDR, only three also included historical responsibility. This could mark an interesting evolution of CBDR away from developing countries invoking it so as to implore developed countries to take greater action and more toward a call for increased action in general, by all.

Indicators emphasized less often in the Talanoa Dialogue

Past efforts, low responsibility, national capabilities, responsibility (historical, current projected) and equity/fairness were emphasized less in TD submissions than they were in NDCs. Many of these indicators have been employed in the past to justify less ambitious action, so not including them furthers the sense that Parties are calling for increased action by all, immediately.

The Arab Group, for instance, did not claim that past efforts justify inaction in their TD submission, as they had done in their NDCs, nor did they mention historical responsibility. The African Group and LDCs both emphasized acting despite low responsibility in NDCs, but neither included low responsibility in TD submissions. Instead, the African Group focused on international support and vulnerabilities, and LDCs used science and vulnerability to appeal to all Parties to take more action. The LDCs also did not mention historical responsibility in TD submissions, which they had emphasized more than average in their NDCs. The LMDC group was the only group of Parties to increase emphasis on responsibility, both historical and current, as compared to the NDCs.

Japan, Norway and Australia chose not to include cost effectiveness in the TD submissions, although they had done so in their NDCs, potentially not wanting to highlight something often used as a reason for not undertaking more domestic effort. The EU and the EIG/Umbrella Group put less emphasis on current and projected responsibility. Since those countries' emissions are decreasing relative to global emissions, this could be a nod toward taking action rather than parsing responsibility. In general, developed and developing countries avoided including indicators that have been used to justify less ambitious action (see Table 12.1).

Table 12.1   Similarities and differences between NDC and TD submissions

NDCs	TD Submissions
Most referenced indicators	
1. National capabilities	1. International support
2. Emissions responsibility	2. Temperature goal
3. Low responsibility	3. Current efforts
4. Past efforts	4. Scientific report
5. Adaptation	5. Other international commitments
Least referenced indicators	
1. CBDR	1. Cost effectiveness
2. Cost effectiveness	2. Carbon intensity
3. Carbon intensity	3. Low responsibility
4. Scientific report	4. Developed countries not doing enough
5. Temperature goal	5. Emissions responsibility

Conclusion

These findings provide nuance to the indicators that are repeatedly used by Parties in discussions of equity, which the IPCC has recognized to include historical responsibility, capability and developmental needs (Fleurbaey et al. 2014: 287). My analysis provides a framework to better understand changes and patterns over time and across negotiating groups with respect to climate justice as the post-Paris climate regime emerges and evolves. For instance, there is a disconnect between the positions that Parties and coalitions have taken in their NDCs and Talanoa Dialogue submissions and the ones that they are most known for championing, as groups, in the UNFCCC negotiations. Specifically, historical responsibility is underemphasized and decoupled from common but differentiated responsibility – which had, in official negotiations, often been invoked by developing Parties as a call for developed Parties to take stronger action. Instead, temperature goals and references to scientific reports are increasingly used to build calls for enhanced ambition and international support.

Could the finding that historical responsibility – long an impasse in climate negotiations – features less prominently in conceptualizations of equity open new doors of potential cooperation and action? Understanding that the historical responsibility approach will likely no longer be the central node around which perceptions of equity are organized highlights the need to ensure that the result of this shift is a stronger and more coherent approach to climate justice, rather than a sense that equity is no longer important. How Parties conceptualize climate justice will affect how they judge global mitigation action and how they shape their subsequent voluntary mitigation and adaptation goals.

Consequently, further research should be devoted to understanding the content of equity claims within NDCs, especially as they are revised in the future. Researchers should conduct further analysis of (1) how equity links with data on the levels of mitigation, adaptation and finance pledged by countries in their NDCs, and on changes in these pledges over time; (2) how decreased inclusion of 'equity' and 'fairness' in Talanoa Dialogue submissions impacts goals pledged in NDCs, and whether this trend holds going forward; and (3) how civil society organizations and UNFCCC negotiators can use this information to best facilitate increased action to realize climate justice. For instance, the finding that developing countries are calling for increased adaptation and international support rather than parsing out more refined conceptions of responsibility might suggest the need to move research away from carbon budget approaches and more toward understanding new zones of potential cooperation, and toward understanding how to increase and coordinate international support. Research on climate justice would do well by focusing much more on these considerations.

Acknowledgement

This chapter builds on research conducted for Swingle (2016).

References

BASIC experts (2011), 'Equitable access to sustainable development: contribution to the body of scientific knowledge', BASIC expert group: Beijing, Brasilia, Cape Town and Mumbai.

CARICOM (2018), 'CARICOM submission on the Talanoa Dialogue', in *UNFCCC's Talanoa Dialogue Platform*, accessed September 2018 at https://unfccc.int/sites/default/files/resource/82_CARICOM%20Submission%20on%20the%20Talanoa%20Dialogue_FINAL%2029%20March%202018.pdf.

China (2018), 'China's inputs on the Talanoa Dialogue', in *UNFCCC's Talanoa Dialogue Platform*, accessed September 2018 at https://unfccc.int/sites/default/files/resource/104_China%C3%A2%E2%82%AC%E2%84%A2s%20inputs%20on%20Talanoa%20Dialogue.pdf.

Civil Society Review (2015), 'Fair shares: a Civil Society equity review of INDCs', accessed May 2016 at http://civilsocietyreview.org/wp-content/uploads/2015/11/CSO_FullReport.pdf.

European Union (2018), 'Views on the preparatory phase of the Talanoa Dialogue', accessed 1 September 2018 at https://unfccc.int/sites/default/files/resource/BG-04-05-EU%20submission%20on%20Talanoa%20dialogue_2.pdf.

Fleurbaey, Marc et al. (2014), 'Sustainable development and equity', in O. Edenhofer, R. Pichs-Madruga, Y. Sokona, E. Farahani, S. Kadner, K. Seyboth, A. Adler et al. (eds), *Climate Change 2014: Mitigation of Climate Change: Contribution of Working Group III to the Fifth Assessment Report of the Intergovernmental Panel on Climate Change*, Cambridge, UK and New York, NY, USA: Cambridge University Press, pp. 287–318.

Gambia (2016), 'Gambia's INDC', in *United Nations Framework Convention on Climate Change*, accessed June 2017 at https://www4.unfccc.int/sites/ndcstaging/PublishedDocuments/Gambia%20First/The%20INDC%20OF%20THE%20GAMBIA.pdf.

German Development Institute (2019), 'NDC explorer', accessed November 2018 at https://klimalog.die-gdi.de/ndc/#NDCExplorer/.

Lange, Andreas, Andreas Löschel, Carsten Vogt and Andreas Ziegler (2010), 'On the self-interested use of equity in international climate negotiations', *European Economic Review*, **54**(3), 359–75.

Rajamani, Lavanya (2012), 'The changing fortunes of differential treatment in the evolution of international environmental law', *International Affairs*, **88**(3), 605–23.

Swingle, Claire (2016), 'Ambition and fairness: understanding equity through intended nationally determined contributions', Williams College thesis.

UNEP (2018), 'The emissions gap report 2018', accessed 1 September 2018 at https://wedocs.unep.org/bitstream/handle/20.500.11822/26879/EGR2018_ESEN.pdf?sequence=10.

UNFCCC secretariat (2014), 'Lima Call for Climate Action (2014)', United Nations Framework Convention on Climate Change, accessed May 2016 at http:// unfccc.int/.

UNFCCC secretariat (2015), 'Negotiating text', in Ad Hoc Working Group on the Durban Platform for Enhanced Action, United Nations Framework Convention on Climate Change, accessed 4 October 2019 at https://unfccc.int/sites/default/files/resource/docs/2015/adp2/eng/01.pdf..

UNFCCC secretariat (2016), 'Aggregate effect of the intended nationally determined contributions: an update', in *United Nations Framework Convention on Climate Change*, accessed 1 September 18 at https://unfccc.int/resource/docs/2016/cop22/eng/02.pdf#page=1.

UNFCCC secretariat (2018a), 'Overview of inputs to the Talanoa Dialogue', in *UNFCCC's Talanoa Dialogue for Climate Ambition*, accessed September 2018 at https://img1.wsimg.com/blobby/go/9fc76f74-a749-4eec-9a06-5907e013dbc9/downloads/1cbos7k3c_792514.pdf.

UNFCCC secretariat (2018b), 'Synthesis to the preparatory phase', in *UNFCCC's Talanoa Dialogue for Climate Ambition*, accessed September 2018 at https://img1.wsimg.com/blobby/go/9fc76f74-a749-4eec-9a06-5907e013dbc9/downloads/1cu4u95lo_238771.pdf.

UNFCCC secretariat (2019), 'NDC registry (interim)', in *United Nations Framework Convention on Climate Change*, accessed October 2018 at https://www4.unfccc.int/sites/NDCStaging/Pages/All.aspx.

13 Responsibility for climate justice: the role of great powers

Sanna Kopra

This chapter contributes to the scholarly field of climate justice by scrutinizing the role of great powers in debating and implementing climate responsibility. The contemporary literature on international environmental governance often emphasizes the role of small states, intergovernmental organizations and non-state actors in promoting ambitious climate change mitigation agendas at the local, regional or global level, often viewing great powers mostly as veto actors blocking attempts at progress in international climate negotiations (cf. Bukovansky et al. 2012, Clark 2011, Kopra 2019a). While this may often be the case in the political realm, it does not mean that we cannot – or should not – expect great powers to shoulder special responsibilities for climate change mitigation and adaptation. As Hedley Bull (2002 [1977]: 200–201) notes, the concept of *great power responsibility* is not a 'description of what great powers actually do' but 'rather a statement of the roles they can, and sometimes do, play that sustain international order' and justice. Therefore, this chapter asks: Can we assume that great powers shoulder more responsibility regarding climate change mitigation than smaller states?

The chapter builds on the English School of International Relations (IR) scholarship, which is a theoretical enquiry rooted in world history, international law and political theory. The English School was developed from the British Committee on the Theory of International Relations, founded in 1959, and especially the works of Hedley Bull, Herbert Butterfield, R. J. Vincent, Adam Watson and Martin Wight. Since Barry Buzan's prominent volume *From International to World Society? English School Theory and the Social Structure of Globalisation* was published in 2004, there has been a notable rise in English School scholarship looking at the social structures of international society. According to Hedley Bull's classic definition, this society 'exists when a group of states, conscious of certain common interests and common values, form a society in the sense that they conceive themselves to be bound by a common set of rules in their relations with one another, and share in the working of common institutions' (Bull 2002 [1977]: 13). In recent years, much less attention has been paid to the normative theorization that was advanced within the English School by Andrew Linklater, Tim Dunne and Nicholas Wheeler, among others, especially in the 1990s.

In addition to its recent focus on societal approaches, the English School is also a normative and practice-guiding theory in a moral sense. As the severe impact of climate change starts to shape the practices of international society, it is high time to renew the English School's interest in moral-philosophical discussions. In order to spur discussion on how international practices *ought to be* in the era of climate change, this chapter studies great powers' role in international climate politics empirically and normatively. It develops a normative framework reasoning why great powers ought to shoulder special climate responsibilities and also analyses how the UN Security Council (the key great power club), as well as the United States and China (the two most powerful states in the world), define great power climate responsibilities in practice.

Great powers in international society

It is a commonplace in IR to argue that great powers not only have special rights in international society but also have special responsibilities in maintaining international order and providing public goods. While materialist theories of IR have mainly focused on the ways that shifts in the balance of power shape international order, social theories are more concerned with how power shifts influence normative settings of international relations. For the English School theory, material capabilities are essential elements of being a great power, but, more important, power is a 'social attribute' which must be placed 'side by side with other quintessentially social concepts such as prestige, authority, and legitimacy' (Hurrell 2007: 39). Thus, the concept of great power responsibility is also inherently social in nature.

In general, the English School assumes that 'even if a state reaches a certain level of material capacity and has a certain national identity, it does not automatically become a great power, but instead needs to be recognised and accepted by other recognised great powers' (Kopra 2019b: 153). Great powers need to have certain material capabilities, but foremost is an *identity* created in the interaction among states. What makes the English School's conception unique is that it maintains that:

> [G]reat powers are powers recognized by others to have, and conceived by their own leaders and peoples to have, certain special rights and duties. Great powers, for example, assert the right, and are accorded the right, to play a part in determining issues that affect the peace and security of the international system as a whole. They accept the duty, and are thought by others to have the duty, of modifying their policies in the light of the managerial responsibilities they bear. (Bull 2002 [1977]: 196)

Due to the anarchic nature of international society, however, great powers' rights and responsibilities are not fully formalized and written into international treaties (Bull 2002 [1977]: 221), apart from Article 24 of the UN Charter (1945), which *appointed* permanent members of the UN Security Council to have 'primary responsibility for the maintenance of international peace and security'. Due to its members' status of legalized hegemony (Simpson 2004: 68), the UN Security Council can be viewed as the key 'great power club' (Bull 2002 [1977]: 194, Wight 1999 [1946]: 42, Kopra 2019a: 70–73) of our times. Hence, it is the members of the UN Security Council that negotiate the 'content' of great power responsibility in time and place. It indeed seems that the notion of great power responsibility is flexible concerning its content and direction: great powers mould it through their discourses and actions at UN negotiations and beyond (Kopra 2019b).

For the English School, the so-called *pluralist–solidarist debate* on the possibility and potential of shared interests, norms, values, rules and institutions in international society (e.g. Bain 2014, 2018) plays a key role in scholarship on the justifications and scope of great power responsibility. While pluralists emphasize the importance of great power management for sustaining international order and security of states, solidarists highlight great powers' responsibility to promote international justice, human security and the well-being of individuals globally.

Taking a very state-centric approach to international relations, pluralism is mainly concerned with interstate order in international society. From their viewpoint, international order constitutes the key means to facilitate peaceful coexistence and other ultimate goals of international society. Therefore, great powers have special managerial responsibility to maintain international order so as to 'ensure that the conditions of international peace and security are upheld' (Jackson 2000: 203). Traditionally, pluralist theoretical and empirical enquiries have focused on great powers' responsibilities in managing their relations with one another in a prudent way in specific situations (Bull 2002 [1977]: 200, Watson 1982: 201), as well as their role in mediation of international conflicts and preservation of the general balance of international society (Watson 1982: 201). Today, however, it is increasingly clear that climate change is a potential source of international conflict, and it poses national security risks around the world (e.g. Barnett 2003). For some states, climate change is even a question of state survival. Small island states in the Asia-Pacific will be lost to rising sea levels caused by climate change in the future. Hence, it is reasonable to argue that climate change is likely to cause a risk to the status quo in the international order. From a pluralist perspective, this means that given their *managerial responsibility* to maintain international peace and security, great powers can be assumed to have a special responsibility to lead international efforts to reduce emissions (cf. Kopra 2019a: 74).

Solidarism is based on cosmopolitan ethics that gives moral priority to the universal rights of individuals over state sovereignty. Thus, it takes individual human beings around the world as moral referent objectives of state responsibility, including great power responsibility (see, for example, Harris 2016). Hence, solidarists underline social attributes of power and responsibility: ideational power is important. As Wheeler (2000: 2) puts it, it is important to '*distinguish between power that is based on relations of domination and force, and power that is legitimate because it is predicated on shared norms*' [italics used in original]. Solidarist scholarship has focused especially on the question of humanitarian intervention and the 'responsibility to protect' (R2P) (e.g. Wheeler 2000). When it comes to great power responsibilities, solidarists underline that great powers have special responsibility to promote international justice. They promote 'universal' liberal ideas that great powers ought to advance in international relations, such as human rights, the rule of law and good governance, as 'new standards of civilization' (Gong 1984). Climate change can be added to the list due to its adverse effects on human security and well-being around the world. From a solidarist point of view, this means that due to their *leadership responsibility* to promote human values and international justice, great powers ought to make serious diplomatic efforts to spur the political will necessary to increase the ambition of climate change mitigation at the global level (cf. Kopra 2019a: 74).

Thus, despite its very limited interest in questions of climate change so far (see Falkner 2012, Falkner and Buzan 2017, Kopra 2019a, 2019b, Palmujoki 2013), the English School theory indicates that great powers can be expected to have special responsibilities in the context of climate change mitigation. The next section investigates the extent to which China, the United States and the UN Security Council acknowledge this responsibility in practice.

Great powers and notions of special climate responsibility

The UN Security Council organized the 'first-ever debate' on the nexus between climate change, energy and security in 2007 – despite China and Russia having challenged the adequacy of the Council for such a debate (United Nations 2007). The president of the Council, the British Foreign Secretary Margaret Beckett, argued, however, that the Council must address the security impact of climate change because the 'Council's responsibility [is] the maintenance of international peace and security, and climate change exacerbated many threats, including conflict and access to energy and food' (United Nations 2007). In line with this, the UN General Assembly (2009a) urged relevant UN organizations to strengthen their efforts to tackle climate change, including its 'possible security implications', and asked the UN Secretary-General to prepare a comprehensive report on the potential security impacts of climate change. In response, the Secretary-General delivered a report which defined climate change as a threat multiplier that could affect security through five channels: vulnerability, development, coping and security, statelessness, and international conflict (UN General Assembly 2009b). Two years later, the UN Security Council (2011) adopted its first-ever statement on the potential security impact of climate change. Yet it made no decision on the use of new environmental peacekeeping forces ('green helmets') in settling conflicts caused by resource scarcity (United Nations 2011). In 2013, 2018 and 2019, the Security Council also discussed climate change but failed to define it as an international security threat, especially because of the resistance from Russia and China (Krause-Jackson 2013, UN Security Council 2018, Pohl and Schalle 2019).

Nevertheless, US President Barack Obama made an explicit link between great power responsibility and climate change in his speech at the UN Climate Change Summit in 2014. After meeting there with Chinese Vice Premier Zhang Gaoli, he said that it was his belief that, 'as the two largest economies and emitters in the world, we have a special responsibility to lead. That's what big nations have to do' (Obama 2014).

A decade prior to Obama's acknowledgement, it had become clear that China was rising to the status of a great power. Concerned about the ramifications for established liberal norms and institutions, the United States called on China to become a 'responsible stakeholder' (Zoellick 2005) and to take 'a role in which a growing economy is joined by growing responsibilities' in international society (White House 2009). As China had surpassed the United States as the biggest carbon dioxide emitter in the world in 2006 (PBL Netherlands Environmental Assessment Agency 2007), climate responsibility was incorporated into the broader calls for China's enhanced great power responsibility. Since the Chinese government did not want to be regarded as an international threat, it developed the concept of 'peaceful development' and started a campaign to pursue a favourable international image (e.g. Deng 2008). On the one hand, the scope of international responsibility remained heavily debated in China (Shambaugh 2013); on the other hand, the state's badly damaged international image in the 2009 Copenhagen climate conference increased the pressure on China to take a more constructive role in international climate negotiations (Kopra 2012). In addition, domestic environmental problems were an important driver for China's growing willingness to address climate change. Over the years, the Chinese government seemed to learn that climate responsibility was a beneficial way to describe China's great power responsibility. In contrast to other attributes of great power responsibility, such as R2P, climate responsibility is not based on the Western liberal ideals. Besides, the fulfilment of climate responsibility was not viewed by the Chinese as hampering the state's overall national interests. Rather, 'greenification' of its economy would support its ongoing structural reforms (Kopra 2019a).

In 2014, China's Special Envoy Zhang Gaoli announced at the UN Climate Summit that 'responding to climate change is what China needs to do to achieve sustainable development at home as well as to fulfil its due international obligation as a responsible major country' (Zhang 2014). President Xi Jinping (2014) also declared at the 2014 APEC that as 'its overall national strength grows, China will be both capable and willing to provide more public goods for the Asia-Pacific and the world'. Although Xi did not explain what he was referring to by 'public goods', his comment hinted that China could be willing to shoulder great climate responsibility as well: clean air is a 'public good'. Notably, a few days later, Presidents Xi and Obama informed the world in their historic joint climate statement that China would stop the growth of its carbon dioxide emissions by around 2030 (White House 2014). The joint statement encouraged international society to believe that an international climate treaty was possible to reach in Paris in the following year, not least due to the acceptance of the United States and China to take great power responsibility to lead international efforts to tackle climate change. The Paris Agreement on climate change was adopted in 2015, and China and the United States were among the first countries to ratify the Paris Agreement in September 2016 – a decision that was made public in a joint press conference indicating once again great power climate responsibility, which probably increased the willingness of other states to ratify it as well. The Paris Agreement entered into force on 4 November 2016, and only a few days later climate sceptic Donald J. Trump was elected president of the United States – an election that ended great power cooperation on climate change at once.

Due to Trump's reluctant approach to international climate politics, and particularly his decision to withdraw the United States from the Paris Agreement, many policy-watchers around the globe now expect China to fulfil the leadership vacuum left by the United States. For China, climate responsibility indeed seems to be an appealing way of defining great power responsibility in the 21st century (Kopra 2019a, 2019b). In practice, however, China has used its increased bargaining power to reintroduce the bifurcation between developed and developing countries in the allocation of international responsibilities – a division abandoned by the Paris Agreement. Presently, China's nationally determined contribution to the Paris Agreement is 'little more than business as usual' (Harris 2017: 102) and 'highly insufficient' (Climate Action Tracker 2018) to prevent dangerous climate change from happening. In short, China has not pledged to reduce its *absolute* emissions but only its *relative* emissions per unit of gross domestic product. Moreover, it has promised to achieve the peak in emissions growth around 2030 without mentioning how much its emissions will increase before the peak. On the one hand, it is fair that China emphasizes its 'development first' principle, given that a big proportion of its population continues to live in poverty. On the other hand, China's climate policies fail to acknowledge that the country's new, rapidly growing affluent class produce as much as – or even more – greenhouse gas emissions than many Europeans and Americans, but those Chinese people are able to 'hide behind China's developing country status' (Harris 2017: 105).

As Harris (2010, 2017) proposes, a principle of common but differentiated responsibilities *among people* (instead of that *among countries*, as in the international climate change agreements) would better capture the global role of affluent Chinese individuals and therefore promote genuine solidarist climate responsibility in international society. Clearly, this is not how China or other great powers define climate responsibility. In addition to the lack of ambition in its domestic climate policies, China has not initiated innovative or effective solutions to increase ambition for emissions reduction at the global level. Hence, it remains very unclear whether China really will be – or even wants to be – an inspirational leader that can truly lead in international climate politics.

Pluralist–solidarist debate and the fulfilment of great power climate responsibility

The present 'great power club' has not put forward notions of great power climate responsibility in a way that is ambitious enough to prevent dangerous climate change from happening. This opens up a critical normative question: How should great powers define their special climate responsibility, and more important, how should they operationalize that responsibility in practice? The English School's pluralist–solidarist debate can help us to address great power responsibilities in the era of climate change. A new English School research agenda looking at great power climate responsibility is needed.

Although the English School in general has its roots in the social theories of IR, its pluralist camp comes quite close to the basic tenets of Realism, especially the tendency to emphasize the material dimensions of power. Since pluralism focuses on the 'is-side' (as opposed to the 'ought-side') of international relations, it is natural that its conception of great power responsibility stresses the importance of the material capabilities of great powers to maintain international order. Military capability is often regarded as a – if not *the* – key attribute of great powers: they are allowed to use armed force, and indeed they are required to do so, if it is needed to maintain international order. Traditionally, coercion has not been viewed as a feasible means of climate change mitigation. However, the role of military power cannot be fully ignored when discussing great power climate responsibility. If climate change proceeds in a disastrous and abrupt manner, it is not very hard to imagine militarization of climate change. Since the UN Security Council has already addressed climate change, it is possible that new climate peacekeeping forces will be established in the future. In line with the Council's mandate, great powers would then play a key role in making decisions as to how and where those forces would be used.

From a pluralist perspective, great powers can fulfil their managerial responsibilities in international society by pursuing their interests with prudence. They should not jeopardize international order but act in accordance with laws and practices sustaining it. Upholding international law is thus an important attribute of great power responsibility: great powers must obey international law in order to maintain international order and legalize their hegemonic status (Simpson 2004). When President Trump decided to withdraw the United States from the Paris Agreement in accordance with the rules of the agreement itself, he did not violate international law per se. Yet Trump's decision was widely criticized around the world, which demonstrates that responsible international actors are expected to participate in the workings of international regimes and organizations (Kopra 2019b: 151–2). Moreover, one reason for the criticism may have been that Trump's hostility towards climate politics is not 'in conformity with a rule, even though that rule is not agreed, not enunciated nor even fully understood' (Bull 2002 [1977]: 216), of great power climate responsibility. The pluralist approach condemns the United States for not fulfilling its great power responsibility because the withdrawal from the Paris Agreement undermines climate change mitigation and thus increases international climate-related security risks that may shake up international order.

Moreover, pluralist ethics also maintains that in line with their material capabilities, great powers bear a special responsibility to solve the problem of climate change. Since efficient emissions reductions are regarded as a key means of stopping climate change, great powers can be expected to implement the biggest share in emissions reduction. In other words, great powers must prepare and implement effective domestic climate change mitigation plans. It is worth noting that the ratification of an international climate agreement is not a precondition to fulfilment of this domestic responsibility, but a state may nevertheless prove to be a responsible member of international society by undertaking ambitious domestic measures to mitigate climate change on a voluntary basis. In that case, climate change mitigation plans presumably support its overall national interests, which is a sufficient justification from the perspective of pluralism.

For the solidarist wing of the English School, however, national interests are not a legitimate reason on which great power responsibility should rest; great power responsibility has to be based on human values and international justice. As other chapters in this volume make very clear, climate change is inherently an issue of human security and international justice. It is a matter of equity and social and distributive justice, and it violates basic human rights, such as the right to life, the right to health and the right to subsistence (see Harris 2016). Given great powers' special responsibility to decrease human suffering, solidarist ethics assumes them to shoulder special responsibility in tackling climate change. Due to the impending human security effects of climate change, it is possible that the scope of R2P will expand in the future. A norm of 'responsibility to protect climate' may emerge, and great powers can be expected to play a central role in that process as well as in implementing that norm in practice. From a solidarist perspective, however, the UN Security Council may not be a legitimate forum to address that norm because, without a reform, the Council is unable to truly promote international justice. In the present form, its membership is based on the post-World War II international order and its military and sanction-based tools are not designed to solve non-traditional security issues or to promote human well-being, for instance. Hence, a new, more representative international forum would be necessary to advance a genuinely solidarist 'green' R2P from climate change.

For many English School scholars, international law is one of the most important institutions of international society. It is constitutive of international society as it captures shared rules of coexistence accepted by members of international society at large. Yet compliance with international law can hardly be seen as an ambitious attribute of great power responsibility because it tends to pronounce only a minimum standard of conduct in international society. Truly responsible members of that society can be expected to do more. Generally, both pluralists and solidarists agree that it is sometimes necessary to violate international law in order to advance the common good of international society. According to pluralist ethics, great powers have a responsibility to act against international law if it is necessary for the maintenance of international order (Bull 2002 [1977]: 138). Moreover, solidarists notice that, like all human constructions, international law may be imperfect and unfair; it may undermine fundamental human values, for instance. In that case, great powers should undertake ambitious diplomatic efforts to develop new, more just international norms and rules.

As for climate law, international climate treaties are not consistent with scientific models of required actions to halt climate change. For instance, states' nationally determined contributions to the Paris Agreement are highly insufficient to reach the goal of limiting the global temperature rise to 2 degrees Celsius (UN Environment 2018), not to mention the goal of 1.5 degrees Celsius that the recent report of the Intergovernmental Panel on Climate Change (2018) regards as the precondition for a safer future. Therefore, solidarists assume responsible great powers should go well beyond the basic requirements of international (climate) law by promoting human values and well-being globally. They should do this on a voluntary basis as well as by taking a leadership role in international negotiations on more ambitious climate mitigation and adaptation policies.

Furthermore, solidarist ethics opens up an intriguing question about the relationship between great power responsibility and sacrifice: Do we expect that responsible great powers should sacrifice their own good for the good of international society? In the context of international climate politics, do we assume that a responsible global leader puts its own national interests aside and commits itself to (economic) sacrifices to make a low-carbon international society possible? Evidently, neither the United States nor China has viewed their great power climate responsibilities in this way. Trump's decision to withdraw the United States from the Paris Agreement clearly dismissed the common good of human beings globally. He focused exclusively on the very narrowly defined national interests of the United States. China's climate policies are not formulated for the sake of climate mitigation per se, but they combine responses to other domestic challenges unrelated to the common good of international society, such as national energy security, domestic legitimacy, a health crisis due to air pollution, and the necessity to reform the 'old' economic growth model (Harris 2013, Kopra 2019a). Hence, the two great powers' notions of (great power) climate responsibility are very pluralist in nature, and pluralist viewpoints of responsibility are not committed to promoting international justice. To enhance international climate justice, it is thus necessary to develop the English School's solidarist notions of great power responsibility as soon as possible.

Conclusion

Climate change will cause serious harm to human security and increase social disparities around the world in the coming decades. It will also likely induce a profound transformation in the workings of international society, including practices of great power management. However, surprisingly few English School scholars have addressed climate change – and when they have, they have largely focused on the empirical and sociological aspects of international climate politics. In the era of climate change, however, the English School should go beyond structural analyses and restore moral-philosophical debate to the heart of its scholarship. With this chapter I have sought to spur normative theorization within the English School by analysing great powers' special responsibility for climate change mitigation. I identified two ethical lines of reasoning for great power climate responsibility. First, a pluralist approach underlines the great powers' managerial responsibility to maintain interstate order. Because climate change undermines that order, great powers have a special responsibility to implement a substantial share of the reduction of global emissions. Second, a solidarist approach highlights great powers' leadership responsibility to promote international justice and human values around the world. Since climate change undermines equity and the well-being of individuals, great powers should undertake efficient climate mitigation actions at home and assume a leadership role in international negotiations to enhance ambition for emissions reductions at the global level.

During Barack Obama's presidency, the United States and China seemed to reach an agreement that – to paraphrase Spiderman – *with great climate power comes great climate responsibility*. Although the UN Security Council has not issued a formal resolution on climate change, the fact that it has discussed climate security several times indicates this development as well. In practice, however, the UN Security Council, the United States and China have not lived up to their special responsibility to respond to climate change. Global greenhouse gas emissions continue to grow at a very worrying pace (e.g. Le Quéré et al. 2018), putting international society on track to face the most severe impacts of climate change. From the English School viewpoint, this means there is an urgent need for a paradigm shift in international security: climate change must be identified as a key security threat in our time. Given their special responsibility for the maintenance of international peace and security, great powers must *immediately* accept and live up to their special climate responsibility. The UN should take a clear stand on climate security, China should take robust action to halt its emissions growth as soon as possible, and the US should renew its leadership role in international climate politics, without delay.

If the next US president fails to do that, China's climate leadership becomes even more critical in the future: a global emissions peak is not possible without China's contribution. There are domestic incentives for China to take a leadership role in international climate politics because such a role would support its domestic reforms, decrease local pollution, improve energy security, and cultivate China's status as a responsible great power in international society. As an autocratic state, however, China does not have very strong prospects for representing itself as an inspirational global leader that manages to facilitate much-needed political ambition to make international society truly 'green'. This, in turn, calls for the renewal of the great power climate responsibility by the United States. Without ambitious great power leadership, international efforts to tackle climate change are likely to remain insufficient and slow. Hence, an urgently pressing question that future research must address is this: how can we encourage great powers to acknowledge and act out their special climate responsibilities? Civil society actors should also engage in this discussion due to their norm-entrepreneur role in international climate politics.

When it comes to the international justice implications of great power climate responsibility, the English School's pluralist and solidarist approaches are likely to have very distinct practical outcomes. The pluralist notion of great power climate responsibility seems to increase the likelihood of the militarization of climate change and is not likely to promote genuine international (climate) justice. In contrast, solidarism, which is based on cosmopolitan ethics, undoubtedly has a lot of potential to promote international climate justice not only amongst states, but also amongst globally affluent individuals (cf. Harris 2010, 2016). Currently, however, solidarism remains a very underdeveloped area within the English School because of doubts concerning whether it is really possible for states to go beyond pluralism in practice. This is lamentable, as climate change can particularly be seen as a showcase for solidarist ethics.

Future research needs to seize the big picture of great power climate responsibility. In addition to the securitization of climate change, what other normative dimensions give reason to expect that great powers might shoulder special responsibilities in international climate politics? With a new research agenda on solidarist climate ethics, the English School has a great deal of potential to enhance international climate justice in general, and the role of great powers in implementing it in particular. For the future of international society, there could not be a more pressing and urgent research agenda than the implementation of great power climate responsibility.

Acknowledgement

This research was funded by the Academy of Finland (project no. 315402).

References

Bain, William (2014), 'The pluralist–solidarist debate in the English School', in Cornelia Navari and Daniel M. Green (eds), *Guide to the English School in International Studies*, Chichester: John Wiley and Sons, pp. 159–69.

Bain, William (2018), 'The pluralist–solidarist debate in the English School', *Oxford Research Encyclopedia of International Studies*, DOI: 10.1093/acrefore/9780190846626.013.342.

Barnett, Jon (2003), 'Security and climate change', *Global Environmental Change*, 13(1), 7–17.

Bukovansky, Mlada, Ian Clark, Robyn Eckersley, Richard Price, Christian Reus-Smit and Nicholas J. Wheeler (2012), *Special Responsibilities. Global Problems and American Power*, New York: Cambridge University Press.

Bull, Hedley (2002 [1977]), *The Anarchical Society. A Study of Order in World Politics*, 3rd edn, Basingstoke: Macmillan Press.

Clark, Ian (2011), *Hegemony in International Society*, New York: Oxford University Press.

Climate Action Tracker (2018), *China*, 30 November, accessed 3 April 2019 at: https://climateactiontracker.org/countries/china/.

Deng, Yong (2008), *China's Struggle for Status. The Realignment of International Relations*, Cambridge: Cambridge University Press.

Falkner, Robert (2012), 'Global environmentalism and the greening of international society', *International Affairs*, 88(3), 503–22.

Falkner, Robert and Barry Buzan (2017), 'The emergence of environmental stewardship as a primary institution of global international society', *European Journal of International Relations*, https://doi.org/10.1177/1354066117741948.

Gong, Gerrit W. (1984), *The Standard of 'Civilization' in International Society*, Oxford: Clarendon Press.

Harris, Paul G. (2010), *World Ethics and Climate Change. From International to Global Justice*, Edinburgh: Edinburgh University Press.

Harris, Paul G. (2013), *What's Wrong with Climate Politics and How to Fix it*, Cambridge: Polity.

Harris, Paul G. (2016), *Global Ethics and Climate Change*, Edinburgh: Edinburgh University Press.

Harris, Paul G. (2017), 'China's Paris pledge on climate change: inadequate and irresponsible', *Journal of Environmental Studies and Sciences*, 7(1), 102–107.

Hurrell, Andrew (2007), *On Global Order: Power, Values, and the Constitution of International Society*, Oxford: Oxford University Press.

Intergovernmental Panel on Climate Change (2018), 'Global warming of 1.5°C. An IPCC special report on the impacts of global warming of 1.5°C above pre-industrial levels and related global greenhouse gas emission pathways, in the context of strengthening the global response to the threat of climate change, sustainable development, and efforts to eradicate poverty', Geneva: World Meteorological Organization.

Jackson, Robert H. (2000), *The Global Covenant*, Oxford: Oxford University Press.

Kopra, Sanna (2012), 'Is China a responsible developing country? Climate change diplomacy and national image building', *Social and Cultural Research Occasional Paper*, accessed 29 May 2017 at: http://digitalcommons.pace.edu/global_asia_journal/13/.

Kopra, Sanna (2019a), *China and Great Power Responsibility for Climate Change*, London, UK and New York, NY, USA: Routledge.

Kopra, Sanna (2019b), 'China, great power management, and climate change: negotiating great power climate responsibility in the UN', in Tonny Brems Knudsen and Cornelia Navari (eds), *International Organization in the Anarchical Society. The Institutional Structure of World Order*, New York: Palgrave Macmillan, pp. 149–73.

Krause-Jackson, Flavia (2013), 'Climate change's links to conflict draws UN attention', *Bloomberg*, 15 February, accessed 24 September 2016 at: http://www.bloomberg.com/news/articles/2013-02-15/climate-change-s-links-to-conflict-draws-un-attention.

Le Quéré, Corinne, Robbie M. Andrew, Pierre Friedlingstein, Stephen Sitch, Judith Hauck, Julia Pongratz, Penelope A. Pickers et al. (2018), 'Global carbon budget 2018', *Earth System Science Data*, 10, 2141–94.

Obama, Barack (2014), 'Remarks by the president at U.N. climate change summit', accessed 6 February 2019 at: https://obamawhitehouse.archives.gov/the-press-office/2014/09/23/remarks-president-un-climate-change-summit.

Palmujoki, Eero (2013), 'Fragmentation and diversification of climate change governance in international society', *International Relations*, 27(2), 180–201.

PBL Netherlands Environmental Assessment Agency (2007), *China Now No. 1 in CO2 Emissions; USA in Second Position*, accessed 3 April 2019 at: https://www.pbl.nl/en/dossiers/Climatechange/Chinanowno1inCO2emissionsUSAinsecondposition.

Pohl, Benjamin and Stella Schalle (2019), 'Security Council debates how climate disasters threaten international peace and security', Climate Diplomacy, 30 January, accessed 3 April 2019 at: https://www.climate-diplomacy.org/news/security-council-climate-disasters-threaten-international-peace-and-stability.

Shambaugh, David (2013), *China Goes Global: The Partial Power*, New York: Oxford University Press.

Simpson, Gerry (2004), *Great Powers and Outlaw States. Unequal Sovereigns in the International Legal Order*, Cambridge: Cambridge University Press.

UN Environment (2018), *Emissions Gap Report 2018*, accessed 4 April 2019 at: https://www.unenvironment.org/resources/emissions-gap-report-2018.

UN General Assembly (2009a), *Climate Change and its Possible Security Implications*, A/RES/63/281, 85th Plenary Meeting, 3 June.

UN General Assembly (2009b), *Climate Change and its Possible Security Implications. Report of the Secretary-General*, A/64/350, 11 September.

United Nations (2007), 'Security Council holds first-ever debate on impact of climate change on peace, security, hearing over 50 speakers', accessed 29 May 2017 at: http://www.un.org/press/en/2007/sc9000.doc.htm.

United Nations (2011), 'Security Council, in statement, says "contextual information" on possible security implications of climate change important when climate impacts drive conflict', accessed 29 May 2017 at: http://www.un.org/press/en/2011/sc10332.doc.htm.

UN Security Council (2011), 'Statement by the President of the Security Council', S/PRST/2011/15, 20 July.

UN Security Council (2018), 'Maintenance of international peace and security. Understanding and addressing climate-related security risks', S/PV.8307, 11 July.

Watson, Adam (1982), *Diplomacy. The Dialogue Between States*, London: Eyre Methuen.

Wheeler, Nicholas J. (2000), *Saving Strangers: Humanitarian Intervention in International Society*, Oxford: Oxford University Press.

White House (2009), 'Joint press statement by President Obama and President Hu of China', accessed 6 February 2019 at: https://obamawhitehouse.archives.gov/the-press-office/joint-press-statement-president-obama-and-president-hu-china.

White House (2014), 'U.S.–China joint announcement on climate change', accessed 6 February 2019 at: https://www.whitehouse.gov/the-press-office/2014/11/11/us-china-joint-announcement-climate-change.

Wight, Martin (1999 [1946]), *Power Politics*, London: Leicester University Press.

Xi, Jinping (2014), 'Seek sustained development and fulfil the Asia-Pacific dream', accessed 29 May 2017 at: http://www.fmprc.gov.cn/mfa_eng/topics_665678/ytjhzzdrsrcldrfzshyjxghd/t1210456.shtml.

Zhang, Gaoli (2014), 'Build consensus and implement actions for a cooperative and win–win global climate governance system', accessed 29 May 2017 at: http://www.fmprc.gov.cn/mfa_eng/zxxx_662805/t1194637.shtml.

Zoellick, Robert B. (2005), 'Whither China: from membership to responsibility? Remarks to national committee on U.S.–China relations', accessed 29 May 2017 at: http://2001-2009.state.gov/s/d/former/zoellick/rem/53682.htm.

Index